The Little Earth Book

by

James Bruges

to Marion

disinformation

Published by The Disinformation Company Ltd.
163 Third Avenue, Suite 108 New York, NY 10003 / Tel.: +1.212.691.1605 / Fax: +1.212.473.8096
www.disinfo.com

First Disinformation® Edition

Library of Congress Control Number: 2003113892

ISBN: 0-9729529-2-6

Printed in Mexico

Distributed by: Consortium Book Sales and Distribution
1045 Westgate Drive, Suite 90 St Paul, MN 55114
Toll Free: +1.800.283.3572 / Local: +1.651.221.9035 / Fax: +1.651.221.0124 / www.cbsd.com

Attention colleges and universities, corporations and other organizations:
Quantity discounts are available on bulk purchases of this book for educational training purposes, fund-raising, or gift giving. Special books, booklets, or book excerpts can also be created to fit your specific needs.
For information contact Marketing Department of The Disinformation Company Ltd.

Written and researched by James Bruges
Editor: Alastair Sawday
Original design: Springboard Design Partnership
Picture research: Sam Duby
Original illustrations: Mark Brierley and David Atkinson
Cover design & Disinformation® edition layout: Ralph Bernardo
First published in Great Britain by Alastair Sawday Publishing Co. Ltd

Disinformation® is a registered trademark of The Disinformation Company Ltd.

Contents

Introduction

Photographs from space show the world shrouded in a thin veil of atmosphere. This gaseous film is the reason why there is life on our planet and not on others; without it the earth would be frozen, pitted and dead. Like Mars.

It took over two billion years for living cells to transform the toxic soup of the atmosphere. Oxygen combined with other elements to foster the rich variety of life that we know today. Gases interacted with bacteria, plankton and plants. Carbon was captured by photosynthesis and laid down in the earth's crust as coal, oil and gas. Complex systems of biodiversity evolved partly with complementary synergy, partly as a balance of competing interests and partly by the erection of barriers between species – all life forms interacting in ways that we will never fully comprehend. Humanity inherited a magical domain, a fascinating, beautiful and healthy world at a comfortable temperature.

But we are digging up much that was laid down and disgorging it back onto the earth's crust and into the atmosphere. We are felling the planet's lungs, the ancient forests. We are causing mass extinctions and reducing biodiversity. We are unsettling the evolved balance of life by creating novel chemicals, gases, bacteria, viruses, plants and animals that have never been part of nature's evolutionary processes, and allowing them to spread over the earth's surface. We are causing global warming on a scale not seen since the end of the Permian period when carbon dioxide emissions from volcanoes started a chain of events that destroyed nearly all life. Earth is running backwards into the sterile toxic chaos from which it started.

Over 1,600 scientists, including four-fifths of all Nobel science prizewinners, issued the famous World Scientists' Warning to Humanity in 1992. It included these comments: "We are fast approaching many of the Earth's limits. Current economic practices that damage the environment cannot continue. Our massive tampering could trigger unpredictable collapse of critical biological systems that are only partly understood. A great change in our stewardship of the Earth and the life on it is required if vast human misery is to be avoided and our global home on this planet is not to be irretrievably mutilated." In the ensuing decade we have not changed course. Already a third of the planet's "natural wealth" has been lost. Human misery on an unprecedented scale may result from the headlong rush of the wealthy for global exploitation. We may be guilty of the ultimate crime against humanity – allowing the earth's support systems to crumble while we enjoy the temporary benefits of an unsustainable lifestyle.

And human society is crumbling. The 9/11 atrocities against the World Trade Center and the Pentagon – symbols of corporate capitalism and military dominance - raised a question mark over the breathtaking project to bring all cultures throughout the world into a single economic and trade empire in which people, everywhere, are defined not by their culture but as consumers, customers and competitors. Society's essential services are being wrested from governments and handed to corporations whose motivation is profit. The global economy is precarious and growth, though impossible on a finite planet, is offered as the only prescription. Civil society is in conflict with institutions that put the rights of money above the rights of society. The policies of these global institutions have failed – the world is littered with collapsed and collapsing states, inequality has risen to obscene levels and commodity prices, on which the poor depend for income, have been driven down remorselessly so that half the world now lives on less than $2 a day. This has caused anger against the West to explode in Central and South America, Africa and Asia. Meanwhile asymmetric warfare, with ultra-sophisticated hardware on one side and terror on the other, could bring unprecedented destruction. Our civilization has taken a wrong turn – but we can find a new path.

There are inspiring examples of people thinking, acting and dreaming concepts for a new direction. Knowledge is not the whole answer – we exploit what we know but defend what we love. The atmosphere is now appreciated as a "global common," and equal carbon-emission allocation rights, within a sustainable total, would give a basic income to everyone, in addition to providing the incentive for all people and countries to reduce emissions. The monetary system could be re-designed to be stable, encourage local market exchange and end currency gambling. Technologies such as hydrogen fuel cells and biomimicry could reduce damage. Trade could operate cooperatively. Poor countries could be encouraged and helped to nurture their infant industries thus adding value within their borders. Our genetic knowledge could strengthen an agriculture that is in harmony with earth's systems, supports a rural community, and gives security to productive small farmers. Local participation, local economies and local democracy could reverse the alienation that is crippling society. Above all, we could return to the United States' unique constitution and to the ideals of its Founders who struggled against corporate tyranny and resolved that government of the people, by the people, and for the people shall not perish from the earth.

But change must be radical. If the scale of remedial action fails to match the scale of the crises, we face catastrophe.

The United States is dominated by a belief in growth, competition, control, confrontation and all the male *yang* qualities. But a quarter of its citizens can now be described as "cultural creatives" with concerned, caring, co-operative, responsible, holistic, female *yin* qualities. It is these Americans who appreciate diversity of culture, who are dreaming up new ways of doing business, new technologies, new management styles and new forms of social organization – creating the potential once again for a New World. The cultural creatives have inherited the qualities of the Founders of this great nation. Let us hope that the planet gives us time to redress the *yin-yang* balance and allow world affairs to move towards justice and sanity.

How to use this book

Dip in and out as you wish. Take one part at a time. Each chapter is short and to the point, however vast the subject, and aims to provoke you both emotionally and intellectually, whether or not you agree with everything said. Many chapters refer you to books though the views expressed here do not necessarily represent the views of those books.

Better still, try reading it all at one or two sittings so that the ideas, held together in your memory, will form an interconnected whole. These global issues must not be viewed in isolation. Our attitude to the environment, trade, militarism, genetics, as well as our underlying philosophy, act on each other. Please do not skip the chapters on economics, hardly the most seductive of subjects, because faulty economics is at the root of most of our problems.

Think how you might influence your representatives. We cannot predict the future with any accuracy and we are not looking for another Utopia but each step can be either towards or away from environmental survival and social justice. We do not have to submit to the faulty dictates of political cliques of whichever party - another world is possible. Also think of ways in which your own life and that of your community could make a difference. As Adlai Stephenson said of Eleanor Roosevelt: **"She would rather light candles than curse the darkness, and her glow has warmed the world."**

"A single tiny island (England) is today keeping the world in chains.
If an entire nation of 300 million took to similar economic
exploitation it would strip the world bare like locusts."
Mohandas Gandhi 1928

Cod

a symbol for our times

'After you, sir...'

The best fishing grounds in the world were off the coast of New England and the prize fish were cod. On Cabot's return from there in 1497 it was reported "the sea is swarming with fish which can be taken not only with the net but in baskets let down with a stone." A hundred years later cod were reported as big as a man and the fishing grounds served the whole of Europe.

Cod swim in large shoals just above the seabed. They have a long life cycle, spawning only when, after four years, they are large enough. When trawl nets were introduced they scooped up all the fish, big and small, and few lasted long enough to spawn.

It was not until the 1980s that the coastal fishermen of New England realized that, thanks to big offshore trawlers, the cod were disappearing, but the US government took no action. So by 1992 the cod were gone. With the fish went the fishermen. Their skills in navigation, gutting, net mending and fish marketing didn't help them. They were also a unique species that will be hard to replace. The tourists who come to enjoy the picturesque sailing ships in Gloucester harbor are now served filleted cod from Russia and fishermen cut the hotel lawns.

Fishermen will now have to wait up to 15 years for the stocks of cod to return, if they ever do. Young cod are no longer migrating to warmer waters to spawn, perhaps because there are no older cod to lead them. Arctic cod, which have no market value, are moving in and they eat Atlantic-cod eggs and larvae. Other species and other predators are upsetting the natural balance and may also prevent rejuvenation of the cod stocks. So the most prolific fishing grounds in the world may now be dead.

It is a different story in Norway where there was also a crisis in the 1980s. Fish stocks were seriously depleted and catches were falling. The government took drastic action that put many fishermen and boat builders out of work, but within three years the stocks were improving and the fishing trade was saved.

Sixty percent of the world's ocean fisheries are now at or near the point at which yields decline, yet governments still provide massive subsidies to their fishing fleets. An international quota system has been introduced but even that has perverse effects: a third of the fish caught are simply thrown back dead, and only the higher value catch retained for the quota.

But the story gets worse. Warmer water is depleting some species. Chemicals from rivers cause plankton blooms that deprive fish of oxygen. Fish farming is a partial solution and is on the increase but it has severe problems too: interbreeding passes defective genes to wild fish, the farms act as hothouses for disease, and intensively used antibiotics and pesticides pollute coastal waters. The farms also need a lot of fish-food – it takes 10 lbs of trawled wild fish (yes, it's made into fish food) from countries like

Peru to feed 2 lbs of farmed Atlantic salmon. Thus Peruvian coastal fishermen are deprived of their staple diet and of their livelihood – to serve the sophisticated tastes of the rich.

Freshwater species have also declined globally. The 1995 catch was 45% lower than that of 1970. Shrimp farms in East Asia are rarely productive for more than five to ten years, after which the land is unusable due to severe pollution; this temporary industry produces cheap shrimp and undercuts the sustainable production of traditional shrimp farmers.

It may be possible to start restoring the life of the oceans if tens of billions of dollars are not given in subsidies for over fishing, if the certification of seafood from well-managed sources is promoted, if coral reefs are effectively protected, if chemical flows from fields to rivers to oceans are checked, if a network of protected marine areas is enforced, if a moratorium is imposed on unregulated fishing in marine ecosystem hot spots and if local communities are given power over their local fisheries.

But fish may be just one example of a faulty worldview. Somehow market fundamentalism, a belief that markets will resolve all problems in the end, has captured the imagination of politicians and economists. It only takes a moment's thought with fish in mind to realize that this is a remarkably silly idea. Cod have been called "the fish that changed the world." This description has a new meaning. They are now an endangered species and a warning to the world.

St Lucia's corals were damaged and the reefs were over-fished. Now half have been put out of bounds. The fish are back in numbers because they have somewhere to breed and tourists have an area of corals to enjoy.

West African fisheries are on the brink of collapse due to over fishing by the European Union.

The world's fishing industry produces $70 billion worth of fish. It receives $54 billion in subsidies.

90% of large predatory and bottom-dwelling fish have been lost from most fishing regions. Only 10% of Caribbean coral reefs are intact.

Declining hourly cod catch in Europe's North Sea

| 1995 | 1996 | 1997 |

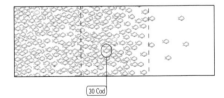

30 Cod

Cod
Mark Kurlansky 1999

Don't Predict!

just be systematic

"Heavier-than-air flying machines are impossible," declared the eminent physicist Lord Kelvin in 1895. The chairman of IBM was wrong too: "I think there is a world market for maybe five computers," said Thomas Watson in 1943. Then scientists predicted that nuclear electricity would be too cheap to meter, and that oil would become scarce by the 1990s; we now extract more than ever. Anyway, we like to think that something will replace oil if it does run out, so why not carry on using it as before? Predictions and their failures can both be misleading.

Instead of guessing at the future we should think systematically about the past and the present - about things we actually know - and **we must live within the rules set by nature.** In a world of finite resource, what is used today will be denied our children – which simple logic avoids the uncertainty of when it might run out. To base our whole economy on gobbling up

non-renewable resources like oil, phosphorus and aquifer water is obviously stupid. There are four principles of sustainability that need to be respected, systematically:

♦ All the toxic minerals like mercury or fossil resources that we dig out of the earth, refine and use will eventually degrade back into the land, water and air and cause cumulative pollution. **We must not extract more than can be permanently contained or safely re-absorbed.**

♠ We live in a balanced ecology that evolved over a period of four billion years. When we introduce new stable molecules, bacteria, viruses, plants, creatures and radioactivity that disturb this balance, they will cause problems somewhere, somehow. **We must not allow these products of society to increase in nature.**

♣ The world's flora, moisture and biodiversity form interconnected natural cycles that support life and a stable atmosphere. **We must not diminish this life-support system.**

♥ The fourth principle concerns equity. The poor, the majority of the world's population, will not endure a distribution of resources that is manifestly unfair. **We must recognize that all people in the world have an equal right to the benefits of nature.**

We must test each human activity systematically and reject it if it offends any of the above principles. There is nothing wooly about sustainability; it is hard-edged, uncompromising, quantifiable and scientifically rigorous. The rules are set by nature, not by man.

The financial world is similar. Commentators constantly predicted that the Dow Jones Index would continue to rise, and they still do. It tumbled but, in the future – who knows? Predictions can be self-fulfilling. The very act of participating in markets influences those markets. The world hung on the words of Alan Greenspan, chairman of the Federal Reserve. If he wanted he could have brought the whole pack of cards tumbling down by making a rash prediction; no wonder his face is so heavily lined. Yet he preferred to remain inscrutable – "if I seem unduly clear to you" he once told a congressman "you must have misunderstood what I said."

Every businessman and gambler would love to predict the future accurately, and every day the financial pages try to help him or her. But the most superficial look at the financial system indicates that it is inherently unjust, unstable and as predictable as the weather. We leave it to the experts who have a vested interest in keeping it inscrutable. The best we can do is keep our fingers crossed in the hope that we may strike lucky.

The economy, like our environmental intervention, has grown exponentially and no one can predict where this will lead. But it should be obvious that we will be in deep trouble if we cannot re-model it to protect the ecosystem and to move, systematically, towards equity and stability.

The Natural Step
Karl-Hendrik Robert 2002

Ozone Layer

a ray of hope

If there were no ozone in the stratosphere there would be little life on earth.

Ozone is a gas, a form of oxygen. It is dispersed in a 20-mile layer of the upper atmosphere where it shields us from ultraviolet (UV) radiation. It is so dilute that if it were all brought down to sea level it would be no thicker than a china plate. Until the ozone layer was in place, nearly two billion years ago, life was limited to primitive forms of algae in the oceans.

In the 1970s scientists worried that within 50 years the ozone layer might reduce by three percent. By the late 1980s this had already happened. But, even worse, by 1996 a hole larger than the US had developed over the Antarctic, another hole was growing over the Arctic, and there was a 20% reduction over much of the world.

UV-B has had a lot of publicity because it causes skin cancers. But it also reduces resistance to AIDS, tuberculosis and herpes, causes cataracts and blindness, reduces the growth and yield of most food crops and kills plankton in the sea (affecting the fish population and also reducing plankton's pivotal role in extracting carbon dioxide from the air). With increased exposure to UV-B the genes in some plants relocate in the genome causing mutations and disrupting the plant's development and fertility. Changes to the DNA of plants and animals could permeate the food chain. Ozone holes are seriously bad news!

How is the ozone layer destroyed?
The major culprits are chlorofluorocarbons (CFCs), though halide and bromide containing substances, widely used in pesticides, are also responsible. CFCs are artificial chemicals that have been widely used in propellants,

refrigerants, insulation, cleaning agents and packaging. They are non-toxic, have no smell and are non-flammable, so for 40 years they were considered benign. There seemed to be no reason to restrict their use. But they offend a fundamental law of sustainability being stable gases not found in nature and somewhere, somehow, they were bound to cause problems. We can now only regret that legislators did not invoke the precautionary principle earlier.

When CFC molecules are exposed to strong UV radiation in the upper atmosphere they are broken up and release chlorine atoms that attack the ozone. A single chlorine atom can destroy many thousands of ozone molecules. If CFC emissions were stopped today the chemical reactions would continue for decades.

Other processes contribute to ozone damage: the destruction of tropical forests, increases in cattle ranching, cattle-feed with a high protein content, artificial fertilization of rice, and the use of chemicals like the methyl bromide in pesticides. It is not a simple matter of cause and effect; there are many interconnections that will be avoided only by working systematically within nature's rules.

There are, however, signs of hope. In 1987, the Montreal Protocol required industrial countries to halt CFC production, and it banned trading in products containing CFCs. **It was based on the "precautionary principle" before all scientists actually agreed that there was a problem.** The conference also set up a fund to help countries change. Although the protocol calls for a global ban by 2006 many industries, particularly in the US, have acted early.

But there are still problems: Russia's economic crisis slowed its program and a black-market in CFCs developed. A refrigerator in every home in China will massively increase its use of CFCs and, if the present US administration succeeds in exempting methyl bromide the agreement could be blown apart. But if these dangers are avoided the ozone layer could be back to normal in 50 years, demonstrating that humanity is capable of rational cooperation.

Protecting the Ozone Layer
Andersen & Sarma

Water
fresh water is running out

Rain is our only sustainable source of fresh water. We ignore this simple fact at our peril.

Fly over the prairie states of America and you will see clusters of dark circles like tiny coins in a desert landscape. Actually, each circle is cultivation, half a mile in diameter, irrigated from a single drop-well and rotating arm. Previously the land had thin grass and wandering herds of buffalo. Now it produces wheat with a higher yield than anywhere in the world. The grain exports from this area are vital for financing America's imports and are a major contributor to feeding the growing world population. It is a miracle of modern agriculture.

Underneath lies the biggest aquifer in the world. Water from melting glaciers seeped into gravel at the end of the last ice age and has been there ever since. It was found in the 1920s and its extraction really got under way in the 1960s. Four to six feet of water is now extracted each year and nature puts back only half an inch. It could last five years or it could last 30 years. No one knows. Farmers believed that this breadbasket was so valuable that the government would provide a massive water project to meet their needs if it ever ran out, like diverting the Mississippi. So they did nothing to economize on the use of water. The future is more likely to be a return of the Dust Bowl that devastated the area in the 1930s.

Most freshwater in the world is in aquifers but extracting them gives us a false sense of security. All over the world the aquifers are being depleted. Some are finite reservoirs that will never be replaced. Some are replenished from surface water but even these have problems. 60% of the nitrate fertilizer we apply when farming remains in the soil and gradually seeps into the groundwater where it is joined by such pollutants as sewage from leaking pipes and chemicals from rusty fuel tanks. Already a major aquifer in China has serious problems with nitrates and some in Britain are laced with benzene. The US and other countries have increasing arsenic pollution.

Irrigation requires organization, so brought the first civilizations into existence. But salt may have destroyed them.

With irrigation, if there is an impervious substrate the water does not drain; salts are drawn to the roots of plants. If the land is drained, however, the salt accumulates in rivers, and evaporation from dam reservoirs concentrates the saltiness. By 1973 the salinity of the Colorado River, where it crossed the border into Mexico's most fertile region, was liquid death to plants.

Water is more valuable than gold – as King Midas found, you can't drink gold. We are as dependent on water to drink as we are on air to breathe but we know even less about groundwater than we know about the weather. How much pollution is already in the ground? How long will it take to get into aquifers? How fast will it spread? How can we extract the pollution? The cost of purifying water is escalating though this is not charged to the polluters – the farmers, the oil companies and the sewerage system. "Let them drink bottled water'" will be the cry when it becomes too expensive to provide clean tap water. And the poor will only drink water if they can pay for it.

Dams are the other source of irrigation. It is strange that the US followed the communist policy of providing virtually free water to its farmers from federal-funded dams, thus encouraging waste – just as in Russia. There was an orgy of construction before dam building finally stopped in the early 1980s.

Virtually all US rivers now have dams. There are 80,000: 50,000 are significant, 2,000 are among the biggest construction projects in the world. But the dams have a finite life as they gradually fill with silt. Already dams are being dismantled. The US is now casting envious eyes on Canada's abundant free-flowing rivers and, when the US gets desperate, its neighbors have much to fear. Do the Canadians want to lose their beautiful valleys, their salmon and their white-water rafting? Under NAFTA rules they may not be able to protect their heritage.

Dams have also seduced India, replacing and destroying centuries-old methods of water-harvesting using bunds, channels and tanka. The Sardar Sarovar dam, at present under construction in earthquake-prone Gujarat on a silt-laden river, has a massive 290 mile concrete irrigation canal that will not be filled for 20 years but is already cracking. Upstream, the Bargi dam, completed in 1990, irrigates only 20,000 acres but flooded over 66,000 acres of fertile land and displaced 115,000 people. Lessons have not been learned.

Global consumption of freshwater is doubling every 20 years. At present:

- 10% is used by people
- 65% is used by industrial agriculture
- 25% is used by industry.

Ancient Civilizations

In 2400 BC, irrigated fields in Sumeria produced over 250 gallons of barley per acre, a respectable yield even by modern standards. By 1700 BC the yield fell to 95 gallons per acre. Soon afterward, crop failures began – and that was the end of Sumeria. Most of the great civilizations that depended on irrigation went the same way - salts came to the surface and destroyed their agriculture.

But Egypt, fertile since the Pharaohs, was an exception. Each year the Nile floods the fields, deposits rich new silt and flushes the surface salts out to sea. Well, it did until 1970 when the Aswan high dam was built. The silt no longer renews the land each year, salts are no longer washed way, bilharzia is rampant, and the delta fishery is declining fast. What's more, the reservoir behind the dam is rapidly filling with the silt that should be fertilizing the fields below.

Cadillac Desert
Mark Reisner 1993

Ecological Footprints

the rich wear big boots

Kerala is beautiful. Canals shaded by coconut palms, markets beside the water, kids playing on the banks. Walk the streets of Trichur and you catch laughing eyes as groups chat and women display their colorful saris. There are plenty of festivals. Then return home and you are struck by how drab and glum everyone looks.

Kerala is one of the poorest states in India. Yet the infant mortality rate in Kerala is lower than in some European countries; life expectancy is 72 years (higher than that of black people in the US); 95% of Keralans over the age of seven can read and write; it has a higher proportion of its population with postgraduate degrees than the US. Importantly, population is stable or falling. It is a matriarchal society but the low birth rate is credited to the high level of female literacy.

What has all this to do with ecological footprints? An "ecological footprint" is the productive land necessary to support people in their lifestyle. An American gets his food, minerals and oil from all over the world and all these things use some of the world's limited productive land. His "footprint" is larger than an Indian's. The world has only 3.7 acres of productive land available for each person but, to support its present patterns of consumption, the world needs 5.7 acres per person. This excessive human footprint is trampling the world's available resources, for example:

 10% of land on which to grow food was lost in the last 30 years.

 the Earth's forest, freshwater and marine environments have reduced by 30% in 30 years.

 a third of all fish species and a quarter of all mammal species are in danger of extinction.

So while the population is increasing, the world's resource base is decreasing at an alarming rate and **if everyone adopted the western lifestyle we would need four earths to support us.** But, as Mark Twain commented, "The problem with land is they stopped making it some time ago."

China has developed a taste for western-style consumerism and plans to expand its economy four-fold in the next 20 years. Its people are encouraged to eat an egg every other day for health and the 1.3 billion chickens will need as much grain as Australia produces. For three extra beers a year it will need Norway's grain. If the Chinese develop Japan's taste for fish they will consume the world's entire catch. The number of cars in China rose 40% in 2002 and its main manufacturer aims to increase output fivefold in the next few years, but there is simply not enough steel or oil in the world for China to match western car use. The United States uses a third of the world's natural resources but China has already overtaken it as the leading consumer of meat, fertilizer, steel and coal.

Back to Kerala. They are not harming the earth; their education, health and longevity are comparable with those of the west; their women have equality; their population is stable. They show that one can have a satisfactory lifestyle with a footprint that is well within the carrying capacity of the world – i.e. sustainable. Perhaps we should be looking at their culture for lessons in how to create a sustainable culture for ourselves.

- When Gandhi was asked for his views on western civilization he replied: "I think it would be a good idea."

- The 20% of the global population living in rich countries consumes 86% of the world's resources.

- Scientists are now predicting that the Amazon could become a desert like the Sahara.

India
Available
America

Plan B
Lester R Brown 2003

An Oxymoron?

sustainable development

In Bali fruit fell from trees and life revolved around music and festival. In Ladakh culture flourished in an extreme climate. In Kenya the Maasai lived in harmony with migrating herds, probably showing continuity from the very origin of our species - Abel, beloved of God, till slain by Cain, the modern farmer and park ranger. China discovered most of our inventions and decided not to use them. In ancient Greece anyone who did not spend much of his time in public debate was dismissed as an "idiot." Islam showed how a classless society was possible, developed philosophy and built the most engagingly beautiful buildings like the Taj and the Alhambra. Cultures grew in response to regional conditions, and excess time and wealth were used for buildings, art, ceremonies and jewelry. For four thousand years of civilization consumption hardly increased but **we are left with evidence of wonderful cultural richness.**

Then on January 20, 1949, in his inauguration speech, President Truman defined most of the world as "underdeveloped areas" where "greater production is the key to prosperity and peace." Set against this was Gandhi's previous warning in 1908 "should India ever resolve to imitate England" he said "it would be the ruin of the nation." Gandhi lost and Truman won.

The majority-nations took up the idea of "development" with enthusiasm. Rich countries cut slices from their cake-of-wealth to feed the poor, but they still believed that the magic cake would grow forever and the rich would never need to reduce its portion. In the end everyone would be satisfied.

Within a few years it became apparent that development was not producing jobs, so "manpower development" became the theme.

Then after 10 years it was realized that hardship persisted so "social development" was the key phrase at conferences, followed by the "basic needs approach" and "equitable development." The one thing that could never be questioned was the concept of development itself.

Eventually it was realized that the magic cake was not growing and might even be shrinking. So "sustainable development" was introduced. There is now an army of ecocrats drawing up papers and attending conferences, busily working out how to do more with less. Light bulbs become more efficient. Cars are shared and oil companies plant trees to absorb their fumes. Cities find all sorts of ways to call themselves sustainable. The ecocrats struggle to bring appropriate technology, "forgiveness" and justice to the poor. Even nature itself has been renamed "natural capital" so that it can be included in the sums. It is all very comforting.

Were four thousand years of culture more nourishing than fifty years of development?

The driving force behind sustainable development is the no pain theory, the promise that we can achieve social justice without changing in any way that might be uncomfortable – or is it just putting off the unpalatable moment of truth when we realize that development has been only on behalf of, and to salve the conscience of, the rich. After 55 years of development over half the world is still desperately poor.

Politicians love grand development gestures. In India they propose to link all the major rivers. The scheme is so expensive it could only be carried out with private capital thus handing India's entire water resource to corporations. And it could be an ecological disaster. Local water-harvesting methods would be more effective at a fraction of the cost and sanitation schemes should be organized locally. To retain fertility, solids should be separated at source, composted and eventually returned to the land; liquids can be filtered through reed beds and returned to the ground water.

Development can be wiped out by human-induced global warming. Following Hurricane Mitch in 1998 the Honduran president commented, "We lost in 72 hours what we have taken more than fifty years to build."

Atmospheric Carbon

the ultimate weapon of mass destruction

Two hundred million years ago, when planet earth was recovering from extinctions at the end of the Permian period, forests covered the land, the atmosphere held a lot of carbon dioxide, the air was warmer than now and the oceans were 200 feet higher. Gradually algae, bacteria, plankton and plants, through photosynthesis, captured carbon from the air and released oxygen. As the plants decayed, coal, gas and oil were laid down in the crust of the earth. The concentration of greenhouse gases reduced, the earth cooled and water levels fell. But the process has been put in reverse for the last 200 years.

Scientists have known about this for a long time. The influence on climate of various gases in the atmosphere was first identified in 1827 by Joseph Fourier, a French mathematician. John Tyndall, an Irish scientist, took on the idea. Then in 1898 Arrhenius, a Swedish scientist, coined the phrase "greenhouse effect" and predicted that if concentrations of carbon dioxide in the atmosphere doubled the global climate would warm by 7-10°F, figures remarkably close to current predictions. Scientists continued to argue for a hundred more years but the connection between human-induced emission of greenhouse gases and climate-change is now universally accepted, except by a few oil industry oddballs.

Greenhouse gases blanket the earth and without them it would be 50°F colder, frozen and lifeless. Water vapor accounts for 60 percent of the effect but changes state as gas, cloud or water. Carbon dioxide accounts for over a quarter. We are in an interglacial period during which, for the last 10,000 years, average global temperatures have been around 60°F, making it possible for humans to cultivate land and build civilizations. Carbon dioxide content during interglacials was higher than normal but never rose above 300 ppmv (parts per million by volume).

uncharted territory
carbon dioxide concentrations
during the last 400,000 years
(showing interglacial periods)
and current predictions
(Vostok ice core data)

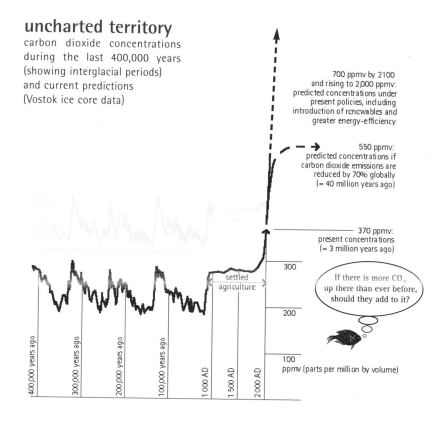

700 ppmv by 2100
and rising to 2,000 ppmv:
predicted concentrations under
present policies, including
introduction of renewables and
greater energy-efficiency

550 ppmv:
predicted concentrations if
carbon dioxide emissions are
reduced by 70% globally
(= 40 million years ago)

370 ppmv:
present concentrations
(= 3 million years ago)

300

settled
agriculture

If there is more CO₂
up there than ever before,
should they add to it?

200

100

ppmv (parts per million by volume)

400,000 years ago
300,000 years ago
200,000 years ago
100,000 years ago
1,000 AD
1,500 AD
2,000 AD

Carbon dioxide concentrations started to rise with the Industrial Revolution when we began to mine and burn coal, releasing the trapped carbon back into the atmosphere. The extraction of oil, which started about 100 years ago, released more carbon. Early last century the concentration of carbon dioxide exceeded the highest levels ever experienced in the 400,000 years for which accurate measurements are available. In the last 50 years concentrations have been increasing even faster, passing 370 ppmv. Even if drastic measures are now taken they will continue to rise to at least 500 ppmv. This is uncharted territory. Global warming may have only just started and we may be too late to stop it. How much will global temperatures change? How high will sea levels rise? It is probable that **we are facing a global emergency on a scale never experienced before.**

Scientists have sophisticated computer models based on what is known. Cosmic rays affect the clouds. Our distance from the sun varies on a 100,000-year cycle, the earth tilts over a period of 40,000 years, it wobbles over a period of 23,000 years and the sun's intensity varies. These periods are so long they may be of only academic interest.

"Positive feedback effects" could be more immediate. The area of snow is reducing. Snow is white and reflects heat but when snow melts, dark ground or water is exposed and absorbs heat causing more warming – so more melting – so more dark surfaces – more warming – more melting. This particular positive feedback effect presents a huge danger. But Hadley Center models show that forests, ground surfaces and the oceans, the natural sinks that absorb carbon dioxide at present, will each reverse with global warming and result in positive feedback effects.

Methane and carbon dioxide, twice the amount contained in all fossil fuels, are trapped in gas hydrates under arctic ice and tundra. If only a part is released warming might get out of control. Polar temperatures are rising five times as fast as general warming and there are widespread reports of tundra melting. Might this cause runaway global warming? The methane threat has not been fed into current computer models. Methane is a greenhouse gas twenty times more effective than carbon dioxide and scientists used to think it was the real problem. They may have been right. When the Intergovernmental Panel on Climate Change (IPCC) was advising the Rio Summit in 1992, a poll of its scientists found that 51 of the 113 questioned believed that runaway global warming was a possibility. Fifteen of these scientists believed it "probable." A Martian would be bewildered at our inaction: "Are the rulers of this planet mad, bad or just totally incompetent?"

Beware of predictions. The IPCC initially predicted a temperature rise of 5°F but a few years later doubled the figure. In June 2003 scientists raised the prediction to 15°F due to the effect of small particles. They are constantly reported as being "surprised" by incidents. In early 2002, for instance, scientists were "astonished" at the collapse of the 1,250 sq. mile Antarctic Larsen B ice shelf in just one month. Perhaps the most sensible comment came from Vladimir Shatalov, a cosmonaut who had the experience of passing through the greenhouse blanket:

"Beyond air there is only emptiness, coldness and darkness. The boundless blue sky, which gives us breath and protects us from the endless black and death, is but an infinitesimally thin film. How dangerous it is to threaten even the smallest part of this gossamer, this conserver of life."

Runaway warming has happened before. There is only one natural system that matches our ability to release carbon back into the atmosphere - volcanoes. The effects of Krakatoa in 1883 could be seen around the world for three years but even that massive eruption made an insignificant contribution to carbon concentrations. To match human emissions you have to go right back to the end of the Permian period. 250 million years ago eruptions in Siberia emitted carbon dioxide that raised global temperatures 10°F, roughly what is predicted as a result of our current emissions. The consequent melting of polar ice and tundra released gas hydrates to cause runaway warming. Almost all life on earth was extinguished and, without plant roots holding it in place, most of the soil was washed into oceans. Evolution stalled for 10 million years while planet earth recovered.

But that was ages ago so could it possibly be a relevant warning for today? Hopefully not. But scientists look for evidence and there is no evidence to suggest a different outcome.

 The Last Hours of Ancient Sunlight
Thom Hartmann 1999

Ostrich Policy

dealing with uncertainty

If one degree of warming is causing such damage, what will ten degrees do?

In September 2002 the Bush administration made greenhouse gas stabilization a national security issue. This followed a report from the National Academy of Sciences (NAS), the leading US scientific body, which gave an even more alarming assessment than the IPCC.

The NAS said that climate changes happen abruptly, not evenly over long periods of time as previously thought but within a few years, maybe a decade or two. Its report says "an abrupt climate change occurs when the climate system is forced to cross some threshold, triggering a new state at a rate determined by the climate system itself and faster than the cause." It illustrated this with the diagram of a seesaw, where a slight shift of weight over the pivot causes the whole structure to tip. The NAS report concludes, "On the basis of the inference from the paleoclimatic record, it is possible that **the projected change will occur not through gradual evolution, proportional to greenhouse gas concentrations, but through abrupt and persistent regime shifts** affecting sub continental or larger regions."

Forget about what might happen to our grandchildren. The climatic seesaw might tip anytime now.

And there is another issue. The sun has increased in luminosity by 30% since life on earth began but the temperature and conditions on earth have remained relatively stable. This remarkable phenomenon gave rise to the Gaia hypothesis that organisms regulate the environment to maintain a temperature that is suitable for life. This might suggest that stability will be maintained whatever we do. James Lovelock, the originator of the science, gives no comfort. He does not consider that the present period is stable. "An interglacial like now," he says "is a period when regulation has temporarily failed and is certainly no time to add more greenhouse gases or deplete biodiversity." Mass human-induced extinctions and changes to the balance of organic life could make things worse and be an added cause of climate change.

Scientists may be worried, but some economists behave like ostriches. Professor William Nordhaus of Yale University, Nobel Prize winner for economics, calculated that the US could afford to spend no more than two percent of GDP to combat global warming because that was the value of the agriculture and forestry sectors which would be affected by climate change. His calculation formed the basis of US negotiations at 1992 Rio World Summit (though, in fact, the US spends nothing like this figure). He ignored the fact that people cannot survive without food, so the monetary value of agriculture would greatly increase if it were seriously harmed by changed climatic conditions. This would be amusing if it did not demonstrate the quality of economic advice on which great nations rely.

Economists then asked "is it worth doing anything?" Cost/benefit studies were carried out. Economists and statisticians made all sorts of assumptions on the extent and effects of climate change in areas that scientists say are impossible to predict. They came up with the suggestion that we should do nothing about global warming, just let it rip and pay for resettling the displaced, it would be cheaper that way. One of the assumptions had to be the cost of death, as a lot of people would die. David Pearce, a leading British economist, insisted that one European or US life is worth 15 Chinese or Indian lives. He was right, of course, within the framework of a discipline that treats people as economic units. He just demonstrated one of the grotesque failings of mainstream economics.

How might we be affected?

A fast temperature change does not allow time for species to adapt. If that occurred, whole biological systems would collapse. If farmers cannot rely on weather patterns the food system is threatened. If the land-based West Antarctic ice sheet were to dislodge, ocean levels would rise by 20 feet - flooding many cities and much of the developed world's infrastructure - including many nuclear reactors. It could be the end of civilization. These are scenarios, not predictions. Almost all predictions are wrong and the actuality will be different. The economists and statisticians mentioned above made the mistake of treating predictions as immutable. Action should not be based on predictions but on systematic thinking that brings our activities into harmony with the processes of a unique planet.

The health of coral reefs has been likened to the miner's canary – if the canary tips off its perch the miner is in trouble – and corals are bleaching around the world. Krill, the staple food of the Antarctic, is depleting. Early springs are leading to a mismatch between the peak food demands of birds and the availability of insects. Some tropical diseases, like malaria and West Nile Virus, are spreading out of their normal areas. Glaciers are receding. Unusual flooding has happened in China, Bangladesh, India, and Africa. Hurricane Mitch killed 18,000 people. But this is only a small sample of the possible effects of climate change. **It is happening due to the rise in temperature of only one Fahrenheit degree. What will happen when this warming is multiplied by 5 or 10?**

The last word can go to the insurance industry, which is the one business that cannot afford to ignore what is happening. With present trends, insured losses will outstrip global GDP in 60 years; we cannot spend all our wealth on disasters, so the crunch will come much earlier even if climate change does not speed up. Already some weather events are not insurable. The industry is pressing governments to DO SOMETHING. Quickly. A first step might be to transfer the whole of the armaments budget to the prevention of carbon emissions – the potential dangers of climate change make terrorism pale into insignificance.

"As this report makes clear, the only questions are how much warming and how soon."
John Knaess, US delegate to the Second World Climate Conference 1990, commenting on the IPCC findings.

Comparative CO_2 emissions:

Boat	30 g/tonne-km
Rail	41 g/tonne-km
Road	207 g/tonne-km
Air	1,206 g/tonne-km

God's Last Offer
Ed Ayres 1999

Europe Cooling

ocean currents

The British rather like the idea of a warmer climate but serious climate change may change their view. Ironically, many scientists now fear that global warming may cause the regional cooling of Europe.

Most scare stories relate to the south. Millions of people in Asia will be flooded due to climate instability that industrialized countries are causing. It is ocean currents that might harm Europe.

Ocean currents have a critical impact on climate. The deep water of the "conveyor belt" in the Atlantic Ocean, which drives the Gulf Stream, takes about a thousand years to circulate at low level around the world, so it was assumed that these currents provided long-term stability. But recently this assumption has been put in doubt. The first shock came recently when currents in the Mediterranean went into reverse in just one year. Then the deep sea current between Greenland and Norway did the same. During part of the last interglacial period average temperatures in Greenland moved up and down by as much as 12°F in a decade, probably due to the switching on and off of Atlantic ocean currents. Although average world temperatures have only risen 1°F in recent years, some arctic weather stations now show warming of 9°F. The thickness of some ice sheets has halved, and their area has reduced by 16%. Strange things are already happening in polar regions.

When ice forms from seawater in the Arctic it leaves the salt behind. The resulting cold, heavy, salty water drops to the ocean floor, creating the "pumps" that drive the Atlantic conveyor. The water in these pumps is now getting less saline, therefore lighter, because less water is freezing. It is also being joined by fresh salt-free water from melting icebergs. And it is getting warmer. If this process continues, a critical point will be reached and the pumps will stop.

The Gulf Stream takes warmth from the tropics to Britain and, without it, the climate would resemble that of Labrador or parts of Siberia. **Its agriculture would be destroyed.** It would then have a crisis as serious as that currently being faced by many countries of the South.

We still do not know enough to predict the precise outcome of an increase in greenhouse gases. Military planners always look at the worst-case scenario and cover it. The opposite is done with climate: the Intergovernmental Panel on Climate Change only gave the "best guess" scenario in its 1990 report, to avoid sensationalist reports. Worst-case scenarios have to be proven before action is taken. By then it may be too late.

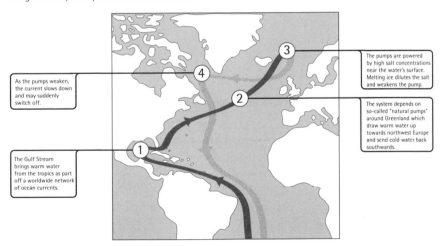

3 The pumps are powered by high salt concentrations near the water's surface. Melting ice dilutes the salt and weakens the pump.

4 As the pumps weaken, the current slows down and may suddenly switch off.

2 The system depends on so-called "natural pumps" around Greenland which draw warm water up towards northwest Europe and send cold water back southwards.

1 The Gulf Stream brings warm water from the tropics as part off a worldwide network of ocean currents.

Rio and Kyoto

let's put it off until tomorrow

At the 1992 Earth Summit in Rio, three requirements for action on climate change were clearly set out:

1. "To achieve **stabilization of greenhouse gas** concentrations in the atmosphere at a level that could prevent dangerous anthropogenic interference with the climate system." (Article 2)

2. The Parties should "protect the climate system for the benefit of humankind on the basis of equity." (Article 3.1)

3. Measures "should be **cost-effective** so as to ensure global benefits at lowest possible cost." (Article 3.3) This hints at emission trading.

No agreement has got near to addressing these requirements. The urgent need to address critical problems with the climate, that brought so many heads of state to Rio, seems to have been forgotten. Due to intense pressure from oil and coal companies it was not until 1995 that the Intergovernmental Panel on Climate Change (IPCC) was able to issue the unanimous, if minimalist, statement that "the balance of evidence suggests a discernible human influence on climate." Ten years after Rio, the "balance of evidence" suggests that there has been no discernible influence on human-caused greenhouse gas emissions despite international conferences, hundreds of select committees and millions of bureaucratic hours. Emissions continue to rise, and global warming has followed the IPCC's worst-case scenario.

At Kyoto, in 1997, a hundred countries, mainly of the North, did agree to reduce their greenhouse gas emissions to 5.2% below 1990 levels within 15 years; but China and India, which have the fastest growth in pollution, were not included. This was totally inadequate for meeting Rio's first requirement to stabilize concentrations in the atmosphere. In addition, each country was given a quota based on its current level of emissions (sometimes called "grandfathering"). This blatantly breached Rio's second requirement, for equity, since the more damage a country had caused in the past, the more it would be allowed to pollute in the future. Nevertheless the Kyoto protocol was an historic first step indicating that the majority of the world wished to cooperate to save humanity from the worst horrors of climate change. The Kyoto protocol was finally signed at Marrakech in October 2001 by 178 countries representing 95.7% of the world's population. Only one country, the United States, walked away from the treaty.

We can expect climate instability and collapsed biological systems. Millions may become environmental migrants; deserts may replace farmland; low-lying agriculture may be engulfed; epidemics may spread. Pacific states are already facing inundation, and their sentiment was well expressed at Kyoto by the President of Nauru: **"The willful destruction, with foreknowledge, of entire countries and cultures represents an unspeakable crime against humanity."** If calamitous climate change occurs, as now seems probable, the slow and inadequate action by the rich minority nations will overshadow all other acts of barbarism in the history of humankind.

Can the Kyoto protocol help? It is a first, hesitant, step in the right direction. It hints at the three Rio principles though they are compromised in ways that gives rise to endless bureaucratic disputes. But it totally fails to meet Rio's challenge or the urgency of the situation.

Emission trading

Although trading pollution sounds bizarre, it was spectacularly successful in the US for reducing sulfur emissions under the tradeable permit scheme of the 1991 Clean Air Act. Most human activity pollutes in some way. Defecation pollutes, but is a part of life's cycle. In moderation burning wood is not harmful. Plants and trees emit carbon dioxide when they die but this helps other plants to grow. Emission rights should be traded only if the aggregate total of these emissions rights is low enough not to cause harm.

Uncertainties

Uncertainties enable bureaucrats to confuse the public, extend negotiations indefinitely and avoid effective action (while being treated as experts and drawing large salaries). The future of the climate is full of uncertainties. Planting forests absorbs CO_2 (good, though only for a time) while their heat absorption in areas that would otherwise be covered in snow causes warming (bad). Soil traps CO_2 but this will be released if the earth warms, as it is expected to do. The effect of small particles (aerosols) is only beginning to be understood but could be dramatic. Methane has not been fed into climate models. Then negotiators come up with "clean development mechanisms," "joint implementation" and other complications to keep the arguments running.

Many scientists are furious with the bureaucrats and say that the problem is too urgent for these quibbles, the only sure way to contain damage is to reduce emissions. John Houghton, former chair of the IPCC working party, describes climate change as a weapon of mass destruction.

Historic Responsibility

Scientists are now able to define the amount of carbon dioxide that has been emitted from each country in the past. This enables allocation of emission rights to be based not only on population but also on a country's total liability for damage, past and present.

Glofs

As Himalayan glaciers melt they form lakes held in by packed ice. These lakes, some of them over a square mile in area and 300 feet deep, are getting bigger. At some stage the ice dams will burst and a wall of water will destroy everything in its path - settlements, bridges and dams. These "glofs" (glacial lake outburst floods) will affect Nepal, Bhutan, India and Pakistan. Will the industrial world, which has grown rich on its profligate use of fossil fuels, undertake massive and urgent remedial action to reduce the levels of the lakes? On present trends their response is to say, "Tough! We need to burn more fossil fuels; how else can we maintain our standard of living?"

The Carbon War
Jeremy Leggett 1999

Contraction & Convergence

one world solution for survival and justice

> "We hold these truths to be self-evident,
> that all men are created equal."
> *Declaration of Independence 1776*

In June 1997, six months before Kyoto, the United States Senate voted unanimously against signing any agreement that did not include all countries within a single timeframe (Byrd Hagel resolution).

At the same time, African, Indian and Chinese delegates said that carbon emission allocations should be proportional to a nation's population. "Emission control standards" said the Chinese delegate, Song Jian, "should be formulated on a per-capita basis. According to the UN Charter everybody is born equal and has inalienable rights to enjoy modern technological civilization."

These two statements are logical and fair. They are perhaps the only issues relating to climate negotiations that really matter. But they did not form part of the Kyoto protocol.

The Global Commons Institute had prepared a framework called Contraction & Convergence that was consistent with both statements and meets the three requirements formulated five years earlier by heads of state at the Rio Earth Summit. Developing nations accepted it as the obvious basis for agreement in 1997 and a senior US negotiator called it "the only game in town." It is also starkly simple.

Contraction: we must obviously contract emissions to a level that will not destabilize the climate. At present scientists think that a 60% contraction on 1990 figures is essential, but further cuts may become necessary.

Convergence: all nations will converge to an equal-per-inhabitant allocation of the right to emit greenhouse gases (GHGs) by an agreed date. Countries can trade their allocations.

Contraction & Convergence is as simple as that. But it has huge implications, in the way that fundamental concepts like "democracy" are simple but have huge implications.

Everyone in the world is dependent on the atmosphere. If it can only absorb a certain amount of GHGs without harm then it is only fair that we should all have an equal ration of emission rights. Not every country nor every person will use exactly its, his or her allocation but, if the allocations can be traded, the frugal will benefit.

A reduction of these gases in the atmosphere is essential for human survival. It is important to note that Contraction & Convergence (C&C)

provides a framework for market incentives to achieve this. Rich nations that emit excessively will have the *incentive* to reduce their emissions as quickly as possible in order to reduce the number of emission-rights they need to buy. Poor nations that do not use their full allocation will have the *incentive* to limit their emissions in order to retain emission-rights for sale. By providing incentives it avoids the pitfalls of a purely regulatory approach or an approach based on horse-trading between nations.

C&C is not totally fair on poor countries that could argue they deserve higher allocations because the rich have already caused so much damage through excessive emissions. They could also argue, with justification, that equal-per-capita allocations should be applied immediately, without a convergence period. But compromise is the essence of politics, and it appears that, at this stage, they might be prepared to compromise if the convergence period is short, say a decade or so. They may change their minds if the rich continue to refuse equal use of a global common.

Equal-per-capita allocations to emit a certain amount of carbon dioxide give every nation a right to something that it can either use itself or sell. And, by extension, the right to this allocation could be the property of each individual as a basic income – in the way that Alaska makes a distribution from its oil royalties to all its citizens, equally.

Contraction & Convergence is therefore not only a framework for reducing carbon emissions to a sustainable level; it is also a market mechanism that will reverse the persistent global tendency to inequality, by establishing a dividend from the value of the atmosphere for all people. It links social justice to the fundamental issue of our threatened future on this planet. Survival linked to justice.

> Why should one person be allowed to pollute more than another?

We are at a critical moment in the earth's history, a time when humanity must choose either to continue on its present disastrous course or to live within the earth's limits and move towards a just and equitable society.

"I call for our nation to join with the world community in solving the challenge of global climate change and work to reduce the emission of greenhouse gases."
Dennis Kucinich

Don't leave it to rich nations. Poor countries tried to establish this approach at Kyoto in 1997 but the principles were sidelined and ignored. History so far shows that rich countries have consistently put their own interests above humanitarian or environmental concerns. If the rich retain control of negotiations they will probably extend the contraction period to reduce benefits to poor countries, even though this threatens their own, as well as others', survival.

China, India, Brazil, (the G21) together with other countries that represent the majority of the global population, should take leadership of C&C immediately, agree on a timeframe for contraction and a rate for convergence, and assert their democratic vote as implied by the Byrd Hagel resolution. Rich countries say they believe in democracy; their claim must now be put to the test. This may be the last chance to avert climatic and social chaos. And this could be the defining decision for the future of humanity.

www.gci.org.uk

DTQs

towards equity in each country

A fair allocation of carbon rights could work locally as well. Everyone, however poor, would then have something of value either to use him-or herself or to sell a basic income.

Diary note: "A good month – only took car out four times and walked to work every day – swipe-card used only a quarter of carbon quota – $350 credited to statement."

Under the Domestic Tradable Quota (DTQ) system you would receive a ration of carbon units each month. The amount of carbon fuel (gasoline or gas) you use determines the amount of carbon dioxide you emit. Your smart-card carbon account would be debited whenever you use fuel with a carbon content, for example at the filling station or when paying the gas bill. There would be a central register at "QuotaCo". **People who use less than their quota could sell their surplus on the open market to people who wish to use more than theirs.**

Once the US undertakes to reduce its emissions to an agreed level, each person would receive an equal allocation. Each state and the federal government should then know that their targets would be met. They would not have to hope and guess that taxes, subsidies and regulations might achieve the desired result. About half the units would be divided equally between everyone in the country and the other half would be auctioned to companies.

Once the carbon units have been issued the market would operate. The family that has a well-insulated house and does not drive a car would have carbon units to sell, while the chief executive who must fly his private jet would need to buy units. The DTQ would not interfere with people's freedom to choose their lifestyle.

Each state could then concentrate on special measures to relieve rural deprivation, to help people insulate their homes and other social benefits. The system would have four results:

1. Everyone would have a commodity that can be sold, providing him or her with a **basic income.** This would reduce the need for arbitrary social measures.

2. There would be a built-in tendency towards **financial equality** because money would flow from those who choose a lavish, carbon-polluting, lifestyle to those who don't.

3. There would be an *incentive* for businesses as well as individuals to **reduce carbon use.** Excess users would reduce their use of carbon in order not to buy allocations. The frugal would wish to keep their use of carbon low so that they have more allocations to sell.

4. The demand for **non-carbon fuel and products** will increase and companies, because of the need to buy units at auction, would reduce carbon use wherever possible.

The quota would initially be based on present emissions of carbon dioxide and gradually reduce, allowing time for businesses and society to adjust. After Kyoto's 12.5% reduction of emissions has been achieved, more reductions will be necessary to bring global emissions down to a sustainable level. Even though the US has not signed up to Kyoto, individual states could introduce the system in order to make their contribution to human survival and to benefit from DTQ's social benefits.

A smart-card system would not suit all countries. In India the obvious choice would be to administer DTQs through the *panchayats* that, in rural areas, are elected from three or four villages; thus the management of carbon allocations and income would be democratically controlled at the lowest level. A system based on the *panchayat* would have the added benefit of encouraging community building while getting a basic income to the poorest.

www.dtqs.org

Tribes: Animals survive and cooperate socially in ways that evolved over eons of time. Wolves have packs, fish have shoals and bees have hives. We humans also have social organization built into our psyche through three million years of evolution – as tribes. Only in the last few thousand years have tribes been replaced by hierarchical organization. Tribes have been exterminated by "civilization" wherever the two come into contact. Free access to land, water and food, enjoyed by tribal society, has been denied to civilized society. Instead of accepting human reality, which was the basis of tribal sustainability, civilization's religions have campaigned for human nature to be changed. Even the word "tribal" is vilified in civilization's languages. Not surprisingly, civilization has been achieved at terrible cost to the majority.

Why did humans evolve in tribes? This question may hold the clue to social exclusion and other pressing 21st-century problems. **Can people be freed from the requirement to drag rocks up the pyramids of the elite?** DTQs could be the first step in providing a basic income that gives people freedom to survive and cooperate within stress-free associations – the new tribes.

With DTQs the government would decide on the number of Carbon Units (CUs) each person would receive. Any units not used could be sold at a price decided by the market. The carbon content of fuels is:

Gasoline	2.3 CU per liter
Diesel	2.4 CU per liter
Natural gas	0.2 CU per kilowatt-hour
Night electricity	0.6 CU per kilowatt-hour
Day electricity	0.7 CU per kilowatt-hour

CU = kilograms of carbon dioxide released

Enlightened self-interest

In 1998, DuPont announced its intention to cut its heat-trapping gas emissions by 65% by 2010. In 2001, the company reached, and surpassed, that goal nine years ahead of schedule. Since 1990 DuPont has succeeded in holding its energy use at these levels. The company attributes a $1.65 billion savings to its program over the past decade.

Beyond Civilization
Daniel Quinn 1999

The Future of Oil

who's fooling whom? – and why?

solar energy captured

solar energy extracted

← extend this line for five miles

Draw a line five miles long to represent the millions of years during which solar energy has been captured and laid down in the earth's crust in the form of coal, gas and oil. Then put a blip in it. That blip represents the time we have taken to extract and use this embodied energy. We are half way through the blip.

These fuels have given us a material standard of living that previous generations could not have imagined in their wildest dreams. Coal gave Britain an Empire and the US gained global power by being the first to exploit oil. Scientific progress would have been stultified without these fuels. But we treat them as a permanent part of the economy. They are not. **Our grandchildren will have to manage without them.**

M. King Hubbert said, in 1956, that oil extraction in the US would peak after 15 years and then decline, never to revive. At the time extraction was relentlessly increasing and oil companies said he was mad. In 1971 US oil production duly peaked and has declined ever since in spite of a desperate hunt supported by the most sophisticated equipment in the world. The US, though it started with 10% of the world's oil (for 4.3% of world's population) is now dependent on imports for two-thirds of its oil consumption. World oil is following the same pattern – the peak will be reached within the next few years.

The history of discovery is well documented and, obviously, more oil cannot be produced than is discovered. US mainland (48 states) oil

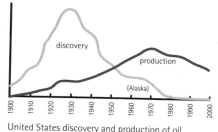

United States discovery and production of oil
(vertical scale as for world x10)

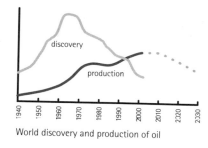

World discovery and production of oil

discoveries peaked in 1930, 40 years before production peaked. World oil discoveries peaked in the 1960s and new discoveries are gradually getting fewer and smaller. New fields will, of course, be found and each will grab the headlines for a time. But they will not change the overall pattern.

The total amount of conventional oil in the world (already used, reserves and yet-to-find) has been about 2,000 Gb (billion barrels) and about half has been extracted. If the amount of production falls and the demand for oil increases, particularly due to the growing Indian and Chinese economies, the price is going to rise – remorselessly.

We are now on a plateau where the amount of oil that can be produced is not much more than the amount in demand. With recession there is adequate production and the price of oil stays relatively low. This enables the world to climb out of recession. But then production cannot keep up with demand. The price rises and the world slips back into recession. This seesaw may last a few years but, well within a decade, the amount of oil we are able to produce will start to drop, never to recover, and recession will deepen. But instability in the Middle East may upset all predictions one way or the other.

President Carter, 1976-80, was aware of the dangers of oil dependence. He said that we must not find ourselves in the position of having to fight wars to seize other people's oil; a couple of decades of transition to renewable energy sources would ensure the stability and future of America without destabilizing the rest of the world. He put in place incentives to achieve solar-heated homes, windmill-powered communities and fuel-alternatives to gasoline. Ronald Reagan, on achieving office, immediately removed Carter's solar panels from the roof of the White House and repealed his tax incentives. The US dependence on oil imports now results in military action wherever there are significant oil reserves.

Saudi Arabia, with the biggest reserves, had a US army presence on its sacred land, a US-trained defense force for its rulers, a US fleet in the Gulf and, under the "Woolsey Plan," it could be broken up and a US protectorate installed in the oil-rich area. The Caspian has significant reserves so, through the Afghan war, the US established permanent military bases in all Central Asian republics - Secretary of State Colin Powell said "America will have a continuing interest and presence in Central Asia of a kind that we could not have dreamed of before." Iraq, with the second biggest reserves, was invaded and subjected to military occupation. Iran is being threatened as it could close the Straight of Hormuz, cutting off the Gulf. Venezuela has the largest reserves outside the Middle East. Through control of world oil the US aims to be in a position to influence the policy of any rival such as China, Japan, Europe or Russia.

Angola has the largest reserves in Africa, but its people do not benefit. Bush, Chirac and oil chiefs were named in the "Angolagate" arms for oil-concessions scandal that devastated the country with civil war. "Commercial confidentiality" keeps dealings opaque, thus making corruption inevitable. Oil executives live in an opulent fenced town with no local contact and fly to offshore rigs - a typical arrangement for foreign exploitation of Africa. The government gets so much income from oil that it has little interest in the domestic economy. Luanda, the capital, is the most expensive city in the world after Tokyo yet two-thirds of its people have no access to clean drinking water and Angola languishes at the bottom of the UN Human Development Index.

For some incomprehensible reason we encourage the wasteful use of oil. Huge subsidies are given to oil-dependent industrial agriculture; air travel is free of tax and no effort is made to reduce the use of cars. A more sensible policy for the US would be to become independent of oil imports by developing non-carbon energy sources.

Hubbert's Peak
Kenneth S. Deffeyes 2001

Hydrogen Economy

the future of energy

"Yes, my friends, I believe that water will one day be employed as fuel, that hydrogen and oxygen which constitute it, used singly or together, will furnish an inexhaustible source of heat and light, of an intensity of which coal is not capable."
Jules Verne, The Mysterious Island, 1874

Iceland could be the first country in the world to emit no greenhouse gases. It may demonstrate what a hydrogen economy looks like.

For 50 years Iceland has been using its plentiful hydro and geothermal electricity to heat buildings and to extract hydrogen from water using an electrolytic process. It is beginning to use offshore wind turbines as well. The process can now be reversed. Hydrogen can be combined with oxygen to become water again and, in the process, drive a fuel cell motor. Hydrogen will be produced at filling stations using renewable electricity sources. First the bus fleet will be powered by fuel cell motors, then private cars and then the fishing fleet. Then Iceland will be totally independent of fossil fuels. Bragi Arnason, who first had the idea in 1978, points to the advantage of Iceland as the pilot country: "It's easy to introduce a new technology in a small society," he says "because if it goes wrong it's less difficult to fix it. Then you take the lessons you've learned here and apply them to larger societies."

But the program in Iceland may not be complete until at least 2040. And that is just one small country that has all the advantages and is well on the way. How long will it take to put a global infrastructure for hydrogen in place? How long to replace machinery with fuel cell motors? There will be a shortage of energy for decades after oil supplies peak and decline.

Hydrogen is the most abundant element in the universe and fuels the sun. Politicians are excited about a hydrogen economy because they can tell the public not to worry, fuel supplies won't be interrupted. And they can tell OPEC to keep oil prices low or we will change to renewable fuels. What's the problem?

The energy you can extract from hydrogen is less than the energy you put into making it - that is the second law of thermodynamics - so using gas or oil to make it is more damaging than using these fuels themselves. Only solar, wind or water flow technologies that produce renewable electricity (or nuclear, but that has other problems) can avoid greenhouse gas emissions. Then there are problems once it is made. Hydrogen reacts with metals making containers brittle, it evaporates even from within a container, it seeps, it has low energy content against volume so has to be stored under pressure, endangering explosion; or it can be liquefied but this needs constant refrigeration. And, though it is difficult to transport, it needs to be available at every filling station and at every airport. Then there are suggestions that extensive use of hydrogen might damage the ozone layer.

But the biggest problem will be the electricity required. To meet our present energy needs with compressed hydrogen for cars and liquified hydrogen for airplanes would require three times the electricity we generate at present. A hydrogen economy must be accompanied by a drastic reduction in our use of energy.

As with all major new technologies research will take time and precaution is necessary. Perhaps the greatest danger is haste but haste may be forced on us when the energy gap looms.

California has got the message. The National Fuel Cell Research Center has been deluged with calls from businesses asking for its advice because of rocketing electricity bills and mains supply failure. The State is offering half a billion dollars. Manufacturers are developing stationary fuel cell equipment, small enough to carry on a truck and capable of serving individual buildings or a local mini-grid. The power will be generated and used locally and there is no need or intention to connect it to the State grid. 21% of California's power will come from wind and photovoltaics using fuel cells within 10 years.

Nuclear Power

too hot to handle

The process starts with uranium, U, the heaviest naturally occurring element. It is also the only one that has 'fissile' content. There are 140 parts of U-238 (238 neutrons) for every one part of U-235 and it is the latter that is fissile – which means its nucleus can be split as nuclear fuel. The uranium is 'enriched' by removing much of the U-238 to achieve a high proportion of the fissile material; this results in a large amount of 'depleted' U-238 uranium (DU), an extremely dense radioactive material. It is this DU that enables shells to pierce armor. Fission takes place when the U-235 nucleus captures (absorbs) a neutron; the nucleus then becomes unstable and splits into two fragments referred to as 'fission products' that are normally radioactive isotopes like strontium-90, caesium-137, iodine-131 etc. The split releases neutrons, which initiate further fission in neighboring U-235 nuclei, and a chain reaction is set up, controlled in a reactor, un-controlled in a bomb. The fission process also releases a few neutrons that are not absorbed so that the mass of the fission products is slightly less than the mass of the original nucleus. It is this tiny difference in mass (m) that releases energy (E), and when multiplied by the speed of light (c) squared (Einstein's famous equation $E = mc^2$), it is a lot of energy. In the chain reaction the fissile uranium undergoes radioactive decay, first to neptunium, then to plutonium. So, when uranium is used in a nuclear reactor plutonium inevitably accumulates as a by-product. Plutonium is another fissile material but it is man-made and not found naturally except in minute traces.

Pluto is the mythical God of Hell and plutonium is named after him - appropriately. It has virtually no peaceful uses apart, possibly, for small amounts in 'breeder' and mixed-fuel thermal reactors. It is among the most toxic substances in existence: inhale a minute particle and you are likely to die of lung cancer; ingest some and you may die of bone or liver cancer. Plutonium comes from 450 civilian nuclear reactors in 29 countries, and from dismantled nuclear weapons in the US and Russia.

Reprocessing plants in just a few countries separate plutonium from reactor fuel. The plutonium is then sent back to the country of origin by ships, air, railways and road. The transport of plutonium is a gift to terrorists. The danger is so great that, even if there is no disaster, democracy in countries with stockpiles may be seriously undermined because of the need for intense and active security – already *habeas corpus* does not apply with Britain's nuclear police. Sellafield, in the UK, has the biggest stockpile in the world. A cupful of plutonium attached to a conventional high explosive could render a city uninhabitable. Whatever the future of nuclear power, reprocessing should be stopped immediately.

Uranium, plutonium and most fission products go through a complex process of decay (including americium which cannot be screened by protective suits) that lasts hundreds of thousands of years while giving off radioactivity.

Extremely low doses of radioactivity can cause cancers and birth defects, where the genome is affected (granite gives off some radioactivity but it undergoes a different decay process that is less dangerous). There is no such thing as a safe level of radiation, so it must not be allowed to increase in the environment. Industry scientists say that the best way to dispose of high-level radioactive waste is to turn it into glass, put it in stainless steel containers, leave it for 50 years to cool and then bury it in stable geological strata. Environmental scientists say that there are no stable geological strata and they recommend perpetual supervised storage above ground. The fact is that **no method for safe disposal of nuclear waste is known.** Then there are growing tons of lower-level waste, and nuclear reactors have to be dismantled and probably covered in earth – to resemble the mounds that Odysseus raised for the honored dead. One wonders what future civilizations will make of these tumuli with their ghostly and ghastly magical properties.

Most nuclear power plants keep waste pools, more vulnerable than the reactor cores, that are vulnerable to sabotage and some of which could equal 70 Chernobyls. Highly active waste is dumping in a desert area where it can leach into aquifers. In the UK 70 tons of plutonium is stored in insecure buildings and a major accident could write off the north of England. If wealthy nations can't be trusted to dispose of the stuff safely, what can one expect of poor nations?

Can we let this continue? The production of radioactive wastes with their dangerous legacy to future generations, the proliferation of nuclear weapons, the exposure of populations to additional ionizing radiation, nuclear accidents, nuclear terrorism, nuclear war, restrictions on civic freedom, and so on, are so frightening and unacceptable that nuclear power should be phased out altogether. Even if we think we can survive the dangers ourselves, can we inflict these problems on our children?

We are trying to answer the wrong question. We should not be asking how to meet humanity's energy demands but "how can humanity live within the constraints of the planet without causing harm?"

Depleted uranium (DU) is a radioactive waste product of nuclear reactors and is given without charge to manufacturers. It is so dense that it penetrates tank armor and is used for bunker-busters and cluster bombs. It burns on impact, releasing radioactive dust particles. US army manuals warn "contamination will make food and water unsafe for consumption." It will be active for the life of the planet causing cancer, kidney disease and birth defects. Half the veterans of the 1991 War, where DU was used, are either dead or sick, their symptoms matching those of visitors to Hiroshima after the atom bomb. Some of the veterans' children are born without arms but with swollen torsos and shrivelled legs, just like many children in Iraq. Much of Afghanistan's water is contaminated by the DU used at Tora Bora. The Balkans are contaminated. Baghdad is a contaminated city. The UN describes DU as a weapon of indiscriminate use that continues to kill after hostilities and therefore infringes the Geneva Convention. The video, "The Invisible War," vividly portrays the horrors of DU and the denials of the Pentagon, but no TV channel will show it. DU must be banned from use in weapons and from any manufacturing process.

In Iraq a mother's first agonizing query about her new baby is "is it normal?"

The Arms Trade

disreputable and deluded

"Few of us can easily surrender our belief that society must somehow make sense. The thought that the state has lost its mind is intolerable, and so the evidence has to be denied."
Arthur Miller

The trade in death and destruction used to be easy to understand. Arms manufacturers were little more than workshops for national armed forces to provide each nation with the means of aggression or defense and to sell weapons to any nation that lacked an industry of its own. Up to a point this is still the public perception. It is wrong. The big arms traders are now multinational corporations.

Is BAE Systems a US or a British company? It started life as British Aerospace but dropped the British for B. It owns former US companies and claims it is "the only European arms company that is truly big in the US and the only US arms company that is truly big in Europe." At the same time BAE wants the UK government to believe that it is British so that Tony Blair will continue to promote it when trotting around the world. But the US accounts for over half of the world's arms expenditure and Congress could, at any time, decide that a company should be excluded for national security reasons. The commercial logic is that BAE should become a fully US company. **The aim of the arms trade, of course, is not to serve any national interest but to maximize shareholder value.**

The government gives massive orders, tax breaks and export guarantees to the arms industry. Since 9/11 it has budgeted more money than can possibly be used effectively. In return, ungratefully, the manufacturers transfer their operations abroad, to Turkey, South Korea, Poland . . . Between 1992 and 1997 the US defense industry laid off 795,000 workers. But why does the government support employment abroad and products of no utility? It would be better to finance narcotics.

The five permanent members of the UN Security Council, who should be aiming to reduce conflict, are the biggest arms traders and all of these governments actively encourage the manufacture and sale of weapons. Most armaments are sold to governments but trade policies have deliberately weakened governments and made it easy for non-state groups to obtain these arms and cause havoc.

The arms trade thrives on conflict. If it can encourage an arms race between two countries, so much the better. In 1997 the US sold weapons worth $270m to Greece and $750m to Turkey when the two countries were threatening war. The principled US embargo on arms to India and Pakistan over nuclear confrontation gave Britain an opportunity – Blair and two top ministers made a special trip promoting arms sales to both sides. The US armed the Vietnamese (against Japan), Iraq (against Iran), Iran (against Iraq), Osama bin Laden (against Russia) and even the Taliban. And the encouragement of arms sales with loans and export guarantees to poor countries is one of the chief reasons for their crippling debt.

US public opinion no longer regards its soldiers' lives as expendable, so they have to be kept at a safe distance from harm. Smart and precision-laser-guided missiles are delivered from highly complex planes traveling at great speed miles away, too far to recognize wedding celebrations, and nine out of ten victims are civilians – not usually western citizens so, presumably, their lives are expendable. These weapons are soon superseded by more sophisticated ones (good for business) and the dud weapons are sold to unsuspecting states, encouraging further cycles of violence.

A language of euphemism has replaced the horrors of war. Those that declare war and those that profit from selling weapons of war are detached from the results of their actions. They don't like you to hear the screams of a girl having her arm amputated without anaesthetic.

How can "decent" people and "civilized" nations engage in such a disreputable activity whose end product is only mutilation, death and destruction?

The arms industry gives generously to both Republican and Democratic party funds.

Weapons of War

war is no longer an option

"Nine out of ten Americans exterminated." This would be a dramatic headline if any television or newspaper survived to carry it. But it actually happened in the 16th century.

In 1527, 170 Spaniards walked into an army of 100,000. Pizarro captured Atahuallpa, the Inca were routed and piles of gold were transferred to Europe. Although the Spaniards' initial success was due to sharp steel weapons, horses, helmets, body armor and deception, the extermination that followed was due to germs. Evolution ensures that organisms develop resistance to harmful bacteria and viruses, but smallpox was new to America and Americans had evolved no resistance. So smallpox, an unintended bio-weapon, decimated Americans.

Most people live in poor countries. The majority of nations, once colonies, became "young nations with a bright future." But now the wealthy minority glowers at what it sees as a risk-prone zone, a breeding ground for crisis and terrorism, suffering from epidemics, over-population, desertification, violence, and corruption. However these majority nations hold the minerals, the crops, the genetic richness and the potential markets that the minority nations covet, and which they plunder.

In the last century, nuclear weapons created apocalyptic dangers that are still with us. In the new century we have new causes for conflict, new targets and new weapons.

● A third of the world's "natural wealth" has been lost due to over-consumption. So scarcity of resources, more than ideological difference, is now the main **cause** of conflict. The US seeks hegemony through its "war on terror" to secure commodities, including oil. Other governments, weakened and corrupted by globalization, vie with non-state groups to control diamonds, drugs and old-growth timber. The free flow of capital between countries allows individuals and groups to amass fortunes and obtain weapons. Water shortages, worsened by corporate exploitation and industrial farming, cause escalating disputes between and even within countries. Land degradation and a changing climate each give rise to mass migration and exacerbate these conflicts. The wealth of the rich, widely seen on mass media, provokes anger and envy in those who lack basic needs.

It has become difficult to locate **targets**. The US and Britain could lob bombs from a safe distance into Iraq, bring it to its knees by causing starvation and disease through sanctions before invading. But an individual like Osama bin Laden, or an intransigent carrying a handbag with unpleasant contents, is more elusive.

Weapons are new. The rich apply the world's wealth to sophisticated weapons technology. Poor states can now get access to missiles carrying radioactive waste. Non-state groups can obtain lethal chemicals. Computer viruses can be sent over the ether. Microbes can be modified by simply adding key genes to close relatives to make strains resembling Ebola or the 1918 flu that killed 40 million people. Thousands of laboratories around the world have the necessary equipment and they can be spread by an infected individual simply walking through a crowded city. Nanobots may be literally able to eat people alive. An individual with military connections sent the packets of anthrax in 2002.

War in an electronic, nuclear, chemical, and biological age is no longer an option. Germs brought the first American civilizations to an end. Ours may end with President Bush's doctrine of domination, Osama bin Laden's terrorism or Blair's warmongering. If we think war, if war is our consciousness, we will achieve war – and face the abyss. Nothing less than a new consciousness is necessary.

The world's military burns a quarter of the world's jet fuel and emits 70% of ozone-depleting CFCs. The US military generates more hazardous waste than the five largest chemical companies combined.

Cluster Bombs

In Afghanistan 70,000 cluster bomblets, the size of baseballs, await children's games. Cluster bombs are sold openly to anyone at arms fairs with an explanation that the bomblets are, in reality, landmines.

It's safer down here

Costa Rica has a better approach

In 1948 José Figueres, President of Costa Rica, disbanded the army. The military budget was given to education and health. The banks, insurance, all utilities and the railways were nationalized. A wealth tax and a social security system were introduced. Women and Caribbean immigrants were given the right to vote. But it was not an easy ride being in the middle of a murderous region wracked by war and outside meddling. The US tried to overthrow Figueres in 1950 and twice attempted to assassinate him. A century-old dispute with Nicaragua over the use of the San Juan river flared up in 1998 but was settled after two years of patient negotiation without resort to arms. Significantly, it is the only country in the region that has not been invaded or used as a base by the US.

Costa Rica is now democratic, relatively rich, and regularly makes the top 50 in the UN Human Development Index. The educated populace makes it an attractive country for investment. But 'economic violence' may yet destroy the dream. Privatizations, following pressure from the IMF, recently caused widespread rioting.

A former President said: "Costa Ricans have cultivated a civilized spirit, a spirit antithetical to militarization and violence, capable of finding peaceful solutions to conflicts and respectful of the rights of others. In the absence of weapons with which to impose an idea, the only weapon left is reason. Today, people such as myself have become fully convinced that **a country that organizes an army becomes its own jailer.**"

Losing Control
Paul Rogers 2001

Tools for Peace

alternatives to violence are possible

Some women of Wajir, Kenya, put an end to clan warfare. The joining test of their group was "if my clan were to kill your relatives, would you still work with me for peace? If you can't say 'yes', don't join the group." After many stormy meetings with the elders, conferences and active involvement with a multitude of groups, peaceful life has been restored. One innovation was the "Rapid Response Team" consisting of elders, youth and women who rushed to intervene wherever violence flared up. An elder commented "most of what was accomplished was done by people with a heart for peace."

To achieve peace we need to think peace. We need peace consciousness from the bottom-up and from the top-down. A Peace Council should replace the UN Security Council, with members like Mandela, Carter, the Dalai Lama and others who have experience of reconciliation. Each national government should have a Peace Department in place of a Defense Department ("security" and "defense" are euphemisms for war). Each country should have peace academies, devoted to disseminating techniques for resolving conflict, in place of war academies. The first duty of these bodies must be to look critically at the motivations of their own governments before trying to intervene elsewhere.

Conflict is natural and can be a stimulus to creativity. The aim must be to understand the causes of conflict, to use conflict as a tool for growth, and to prevent it slipping into violence.

Violence leads to violence. But the cycle can be broken. An atrocity inevitably induces horror, then grief and then anger. Left to itself, the anger leads to hatred and revenge; further atrocities will set the cycle of violence rolling faster.

Intervention can channel anger into seeking justice through law rather than retaliation. This is better, but still does little to prevent a repetition of the original atrocity.

Intervention can be taken further by analyzing the causes of conflict with a view to removing them and bringing reconciliation. This requires a combination of determined, powerful leadership, imaginative action, and adhering resolutely to key principles. It is a process that can be studied and applied systematically. And it is economical. NATO countries spend 215,000 times more on "defense" as on the main regional organization devoted to conflict resolution (OSCE).

The Oxford Research Group has collected stories of many individuals, groups and organizations using techniques for alternatives to violence. In 1992, when militant Hindus destroyed the Babri mosque and violence spread throughout India, Lucknow avoided bloodshed because school pupils, following the teachings of Gandhi, got religious leaders to work together, and took to the streets in their thousands with loudspeakers appealing directly to the people to refrain from violence, carrying posters like "The name of God is both Hindu and Muslim." In El Salvador, some businessmen started a "guns for goods" scheme, offering a $100 voucher for goods in local shops in return for each weapon turned in. By the end of the second week hundreds of weapons had been collected. They then called on the President for help, and in a short time 10,000 weapons were collected. In Belgrade in 1998 a few hundred students were trained in techniques of non-violent action. They in turn trained 20,000 election monitors, two for every polling station, so that when the elections eventually took place, it proved impossible for Milosevic to rig the results. Some incidents may seem insignificant in relation to global conflict, but politicians know how to make mountains out of molehills. Peace Direct, a new charity, aims to apply these techniques in the field.

There is a tendency to wait too long before intervention so that the use of hardware becomes inevitable. There are usually reliable signs of conflict: the right to vote, or to speak a language or to practice a religion may be threatened; theft or diversion of resources; occupation of territory; arms build-up; warlords making threats. If recognized early enough conflict can often be resolved through dialogue, mediation, bridge building, promotion of democracy, protection of human rights, support for indigenous dispute resolution, election monitoring, confidence building and restorative justice. These tools need to be more readily available than weapons. They are tools that can be studied and applied systematically. And, learning from the women of Wajir, they can be rushed to the site at the first signs of conflict.

Signs of hope: The British government recently allocated £110 million for conflict resolution, and there are 50 institutes in Britain studying small-scale non-violent intervention, such as at Sandhurst and Bradford Peace Studies

"Religious differences do not cause conflict. Religion comes in only once the lines of conflict have been drawn." *Zaki Badawi*

In a scientific conference a delegate held up a child-size talcum powder tin and explained that if this contained TCCD, used by the US in Vietnam, and was dropped into the water supply of a city the size of New York it could kill the entire population.

In 2002 the world spent $794 billion on armaments and $1 billion on war prevention.

"An eye for an eye leaves everyone blind."
Martin Luther King Jr.

War Prevention Works
Dylan Matthews 2001

Inequality

a bequest from the 20th century

In spite of all the benefits of science and technology we have entered the new century with half the world living on less than $2 a day.

A perverse global economy and trade framework controlled by the rich have systematically transferred commodities – old growth timber, minerals and crops – from poor countries to rich countries. In addition to this transfer of *real* wealth the poor are said to owe the rich piles of money as well – "if you don't pay up" say the rich "the global economy will crash." Poor countries, for some reason, sell their dwindling assets and cut social spending for those in desperate need in order to pay.

Four-fifths of the world's population now has to make do with only 14% of the world's wealth; the rest goes to support the living standards of the rich minority.

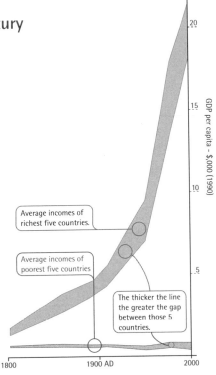

GDP per capita - $,000 (1990)

Average incomes of richest five countries.

Average incomes of poorest five countries

The thicker the line the greater the gap between those 5 countries.

1800   1900 AD   2000

During the last four centuries, Europeans believed it was their God-given mission to exploit and/or settle the rest of the world, taking resources and land from Americans, Africans, Indians, and Australians by force. Military colonization is now frowned upon but economic inequality achieves the same results. The world's money is concentrated in just a few hands. Rich nations and corporations can buy politicians, buy scientists, manipulate public opinion and destroy opposition within their own country or abroad using only their petty cash. A tiny slice of their accumulated wealth is unimaginable riches to an African village, town or even country. The use of force to extract resources is now only necessary in exceptional circumstances.

Even within wealthy countries the benefits of accumulated wealth are not shared. The Human Poverty Index 1998 for industrial countries showed that the US had more people who were functionally illiterate (20.7%) than the proportion in any other industrialized country, the highest proportion of population below the income poverty line (19.1%) and the greatest number of people who did not expect to reach the age of sixty (13%). Ten percent of the US population depended on private charity for food, and 44 million had no health coverage. In most countries inequality is increasing not reducing as increased prosperity benefits only the rich elite, often leaving the poor even worse off. Some African countries, like Angola, with high GDP and massive natural wealth, suffer the worst poverty. **The United Nations predicts that a third of the world's population will be slum dwellers within thirty years.**

There must be something structurally wrong with a world economic system that allows these inequalities to increase. Yet economists say that further economic growth will relieve poverty. And politicians claim that new initiatives on debt, aid and global trade can buck the imperatives of an economic system that, by its fundamental laws, transfers wealth from the poor to the rich.

- The top 1% of households in the US has more wealth than the entire bottom 95%.

- "The globalization of the world economy will continue, with a widening gap between the "haves" and the "have nots."
  Vision for 2020, US Space Command

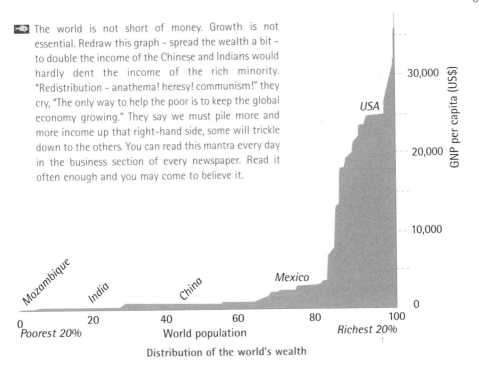

The world is not short of money. Growth is not essential. Redraw this graph - spread the wealth a bit - to double the income of the Chinese and Indians would hardly dent the income of the rich minority. "Redistribution - anathema! heresy! communism!" they cry, "The only way to help the poor is to keep the global economy growing." They say we must pile more and more income up that right-hand side, some will trickle down to the others. You can read this mantra every day in the business section of every newspaper. Read it often enough and you may come to believe it.

GNP per capita (US$)

30,000

USA

20,000

10,000

0

Mozambique

India

China

Mexico

0   20   40   60   80   100

Poorest 20%   World population   Richest 20%

Distribution of the world's wealth

In 1976 Switzerland was 50 times richer than Mozambique. In 1997 it was 500 times richer.

Human Development Report
UNDP 2003

Usury & Greed

religious perversions

*If you advance money to any
poor man amongst my people, you are
not to act like a moneylender; you must
not exact interest from him.*
Exodus 22.25

Usury and greed used to be sins but now they are the twin towers of our economic life. Interest-on-debt, the modern phrase for usury, is how our economy works. Greed - encouragement of the desire for more, more and yet more - is the motor for economic activity.

From the Pentateuch onwards, Jewish and Christian writers condemned usury. The reason is fairly obvious to anyone prepared to free her mind from platitudes about communism. Look at it like this. If you have spare money you can lend it to someone who desperately needs money. She has to work and pay back not only the money lent but some more as well;

meanwhile you can watch your money grow and lend more to someone else. With interest, money flows from the poor person to the rich person, from the poor country to the rich country. **Interest is responsible for the constant growth of inequality.**

If the interest rate is 5%, every company that borrows has to be 5% bigger next year than this year or go bust. With compound interest it has to be twice as big in 14 years, four times the size in 28 years. **Global resources are consumed at an increasing rate because of interest.** Consumption and economic growth are not exactly equal; a fat cat director can only stuff a certain amount of food and drink into his stomach, but he can use his loot in wasteful ways like taking a holiday in space. So interest requires constantly consuming more of the earth's resources. But each country is finite so it cannot grow unless it loots other countries. The world is finite also, so there is a limit to consumption. Interest, therefore, causes wars and is destroying the planet.

Leviticus sets out a benign economy where interest must not be charged, where debt would not accumulate but be cancelled at the Jubilee every fifty years, where those that fail must be supported, where surplus is used for the purpose of allowing the land to rest and recuperate every seventh year.

But Christians ignore their scriptures. Usury and greed are so ingrained in our culture that we no longer question them. Value, which used to refer to the quality of social, intellectual and spiritual life, is measured only by money in the new dogma. The rich minority nations are forcing this pernicious belief-system on all countries and cultures throughout the world, whether or not they wish to adopt them.

Not all cultures have the same values. Islam still does not allow interest to be charged on debt and one of the five pillars of Islam is *zacat*, the religious duty to distribute wealth to the poor. Muslims have good reason for

their fundamental, often passionately held, objection to the imposition of usury and greed as a value system. Buddhists reject acquisitive values. For Hindus, business is just one of the four stages of life. In all religions, value is measured by qualities other than money.

Indigenous cultures work on a cooperative approach. To them, wealth is the tribe and individual windfalls are shared with anyone in need. Being a member of the tribe gives you security in your old age so **there is no need to accumulate money or consume more than is necessary.** These tribes have lasted tens of thousands of years without harm to their environment. In contrast, the TRIPs agreement on patents makes sharing and co-operation a crime.

Culture and religion, through centuries of civilization, explored many ways of understanding and defining the concept of value. Usury and greed were not among them.

Of all vile things current on earth,
none is so vile as money.
Antigone, Sophocles 441 BC

Economic Growth

and the second law of thermodynamics

Frederick Soddy, FRS (1877-1956) was a pioneer atomic scientist who received the Nobel Prize in 1921. He was deeply concerned about atomic power: "If the discovery were made tomorrow," he wrote in 1926, "there is not a nation that would not throw itself heart and soul into the task of applying it to war." The Royal Society ridiculed him for criticizing new advances in science. He in turn criticized the Royal Society's comfortable view that scientists have no responsibility for the uses to which science is put.

Soddy considered that the economic system contained built-in elements for conflict and the destruction of Nature once science gave the power, so he applied scientific thought to economics. He was then ridiculed by economists.

Try this: A sack of grain is real and can be considered as wealth. But it reduces. It rots or is eaten by weevils and gradually turns to dust – that is the second law of thermodynamics. But if you talk about minus one sack of grain you are using a mathematical concept: it is virtual (i.e. conceptual) wealth, or debt, and it can grow *ad infinitum*. Soddy said that **the ruling passion of economists and politicians is to convert wealth that perishes into Virtual Wealth (debt) that continuously grows.**

So economists have turned the world on its head: you can store their money and it grows but if you store something real, like a coat, it decays and gradually loses value. They insist on making the real world of matter which decays conform to the purely mathematical concept of growth and compound interest. This is logical nonsense that can only lead to catastrophe in nature and the breakdown of society. To illustrate the absurdity of compound interest Soddy pointed out that if

Christ had invested £1 in our economic system it would now be worth £1,000 billion billion billion billion billion, or the world's weight in gold. He said in the 1920s that the result of this non-scientific thinking must inevitably lead to debt cancellation, revolution or war.

He referred to development since the Industrial Revolution as the flamboyant period of history, when humans are using up the capital stock of coal. This can only be a passing phase, after which we must live by sunshine – which has a maximum rate of flow. Therefore there is a limit to growth. Looking back after 75 years, has he been proved right?

Answer A:

Economists might say: "He was an amateur who knew little about our discipline. There has been steady economic growth, wealth has multiplied and our standard of living has increased out of all recognition; our scientific achievements have been spectacular and the whole world is now open to markets, communication and tourism. Through our wealth we have gained the knowledge to deal with whatever problems of health and the environment that may arise. There are, admittedly, still pockets of need but the way to fill them is to ensure further economic growth." Is he saying that all our politicians, economists and businessmen are misguided?

Answer B:

Soddy, if still alive, might now reply: "Yes. The Wealth of Nature, real wealth, has been eroded. All our gadgets await the ravages of entropy. We have already passed the limit of growth and further economic growth may remove nature's life-support systems entirely. Individual problems with health and the environment will be swamped by general degradation. Virtual wealth, concentrated in a few private hands in only a small part of the world, has now reached levels that I could not have imagined and is leading to conflict and the breakdown of society. Economists have created a make-believe world that has nothing to do with the real world. To aim at further economic growth is suicidal."

Which is closer to the truth? Take your pick.

Beyond Growth
Herman Daly 1996

Making Money

it's really quite simple

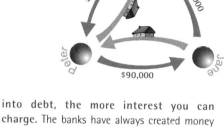

You start with $90,000 and call yourself a bank. You lend it to Peter who wants to buy a house. Peter buys the house from Jane, and Jane puts the $90,000 into your bank for safekeeping. You now hold $180,000 ($90K in cash plus $90K secured on a house).

You can now do the same with Michael and you hold $270,000 (the original $90k plus $180k in collateral). Soon you hold a million, most of it in collateral on houses on which you charge interest. You can now attract deposits at low interest to lend to others at high interest. Or you can invest your notional wealth and make money that way. With enough gall you could have started without even the original sum because they are only computer transactions anyway. **The more money you create by putting others into debt, the more interest you can charge.** The banks have always created money but this system only really took off with deregulation twenty years ago.

Governments make the notes and coins we use but these total less than 5% of the actual money in circulation. Banks and building societies have created the rest in the way described above as computer entries, not notes and coins. The money we use daily was created by private institutions, is owned by them and we pay them for using it, not once but year after year. Banks must maintain 10% 'capital adequacy' so they put aside that amount from net interest; it does not inhibit their ability to create yet more capital, it just puts up the cost of borrowing money and keeps them a nice little reserve.

The Federal Reserve tries to control the money supply by putting interest rates up or down. When they are down people borrow more and the money supply rises, then inflation threatens and interest rates go up – people lose their houses, and businesses go bust. This is a subtle form of torture that governments inflict on us.

Businesses now pay on average 28% of their income to service debt; the average house owners (two salaries) pay a third of theirs to service their mortgages. This leads to intense competition between companies to advertise and produce cheap, short-lived goods, and for consumers to struggle to remain in employment so that they can pay their mortgages. Western societies are aggressively competitive not because it is in our nature but because we must be to survive in a world where nearly half of all wealth is creamed off by the financial sector and the weak go to the wall. We have no choice.

If we work harder and the nation gets wealthier, will our indebtedness reduce? The sad fact is that the reverse happens. In Britain 20% of income services mortgages; the figure is 33% in the wealthier US. In Japan generational mortgages which extend liability to your children have been introduced. No wonder the Americans and Japanese work all hours of the day. Stress is so great in Japan that 10,000 middle-aged men committed suicide in 1998.

Lord Josiah Stamp, former director of the Bank of England, predicted this development in 1937: "The modern banking system manufactures money out of nothing," he said. "The process is perhaps the most astounding piece of sleight of hand that was ever invented."

They've made too much money. US and UK credit and other "promises to pay" had reached ten times the level of real income (GDP) by mid 2003. Japan's economy crashed when this ratio reached nine in their country in 1990.

"To say the financial system is crazy is an understatement. The situation defies description, and it beggars belief that no one involved in the operation takes a long hard look at what they are doing and just bursts out laughing at the innate insanity of the whole process."
Michael Rowbotham

Banks

they've got you by the jugular

Our money system is the only one we know. People with money can watch it grow on Wall Street while they do nothing – it did so for 50 years and they assume that it will go on doing so. Others pin their hopes on the lottery. Others earn an honest crust and assume that they are being treated fairly. But there have been plenty of other systems. While things are going well, however, any writer criticizing the present system is deeply unpopular.

Americans are admired for their facility with money and commerce, so it is surprising that the most damning criticism of the system came from two of their most revered and intelligent presidents. Thomas Jefferson (1801-09) was alarmed at the development of the banking system. "If the American people ever allow the banks to control the issuance of their currency," he said, "they will deprive the people of all property. I sincerely believe that the banking institutions having the power of money are more dangerous to liberty than standing armies."

Abraham Lincoln (1861-65) forged America into one nation. Two main causes of Civil War had been slavery and the financial stranglehold exerted by northern banks over plantations in southern states. Reconciliation was a truly great achievement. Had Lincoln lived longer he might have left an even greater legacy. He wrote a paper called "Monetary Policy" which might have led to the end of conventional banking and money power in the US, and the world might have followed suit. A major *causus belli* would have been removed and the bloody history of our age would have been very different. He was assassinated shortly after publishing this document and it was quietly buried.

Lincoln's Monetary Policy is a masterpiece of clarity; this is a key passage: "the government should not borrow capital at interest as a means of financing government work and public enterprise. The government should create, issue and circulate all the currency and credit needed to satisfy the spending power of the government and the buying power of consumers." Lincoln's suggestion that money should relate to things that need to be bought and sold seems fairly obvious but the banks are playing a very profitable game and such suggestions would spoil the fun.

Jefferson and Lincoln's warnings have come true. The loss of liberty, when poor countries are forced to make structural adjustments that harm their social services, has been dictated and enforced by economic, not military, force. The World Bank and the IMF ensure that the requirements of banking institutions are sacrosanct. Within the US the power of money has made it impossible for communities to resist corporate dominance in all their affairs. And the need for government to borrow money at interest from private institutions has made the US dependent on investment from abroad – investment that could be withdrawn should alternative assets be identified, leading to a crash of the US and probably also of the global economy.

The money supply is now three times greater than the value of goods and services available to be bought or sold and **the main commerce in the world is currency speculation**, untaxed, as $1,500 billion flows around the world every day. All the money that exists attracts compound interest, which is nice for the banks but plunges individuals and governments into debt. Many of the highest paid people in the world are the perpetrators of this destructive and unstable game. It is a game that can destroy nations, destroy even banks that get it wrong, and makes pawns of us all.

So it's risky. Spread the risk. Collateral-debt-obligations - derivatives - you and I don't know what they're playing at; actually nor do they. But the amount outstanding is four times the income of the world. When the music stops it will all come crashing down on our heads.

Interest-free Banking

currencies of the future

The JAK Bank in Sweden has over 20,000 customers and they don't pay or charge interest; they only pay administration charges.

The trick is in savings points. These are not just the amount of money in your account but also the time it is there, so savings points are measured as dollar-months (krona-months in Sweden). Normally you can only borrow up to the number of dollar-months that you have managed to save. But you can borrow up to eight times more if you continue to save while paying back the loan. The dollar-months you save must match the dollar-months you pay back; in this way you build up a nest-egg that cannot be withdrawn until the loan is paid off. **This balance of loans and savings insulates both the JAK bank and its customers from the crises of the money market.** You are not affected by the national interest rate.

The JAK movement started in Denmark in the 1920s and wherever JAK banks were established they increased prosperity in the area. Official banks took every opportunity to crush them because the JAK movement threatens the amazing concession society gives to banks to create money and cream off interest. At last the Swedish JAK bank has found a formula that is difficult for the establishment to stop. And it is spreading.

What about the bigger picture? – After all, this book has railed against interest. Bernard Lietaer, who has been described as the world's top currency trader, suggests that a currency, not just with zero interest, but with negative interest, is necessary and that it might be welcomed by big business.

Our interest-bearing currency forces companies into short-term investment: If you don't need a product till next year don't make it this year – money grows faster than inflation. Expenditure in 10 years costs a third at today's value. This is called discounted cash flow and is a technique used by all financial planners. If a company decides to be conscientious by investing for the future an asset-stripping predator will gobble it up. Currency growth leads to currency speculation, not to investment in real assets.

However, if a currency is constantly losing value, what is called a "demurrage currency," the reverse applies. For example, if you are convinced that solar photovoltaics or fuel cells will have a big market in ten years' time, the sooner you build a factory the better. It will cost a lot more to do so after a few years because your money will have dropped in value. So a demurrage currency is good for business. It favors long term planning and it encourages people to maintain their property. And it cannot be used for gambling.

The money system used to serve the trading needs of big companies. But now less than five percent of trade relates to the real world of goods and services; the rest is currency speculation, growing at 15% a year and not even taxed! (The Tobin tax has been proposed but a demurrage currency would make this unnecessary). Countries, corporations and banks are helpless against this gambling. It requires instability, otherwise there can be no profit in the game. Currency speculation is like an out-of-control cancer at the heart of our system. Not surprisingly, the big companies don't like trading with such an unstable commodity, so a quarter of world trade has gone back to barter. Pepsi, for example, takes its profits from Russia as vodka; France builds nuclear power stations in exchange for oil. This return to barter is surely an admission of defeat for flat earth economists and it is time they started thinking about mechanisms that serve the real world.

Demurrage currencies have served prosperous civilizations well in the past – the Ming dynasty, ancient Egypt, 10th to 13th-century Europe (the age of cathedrals) – and it is now quite possible that companies will turn to the stability of a demurrage currency to escape the merry-go-round of the speculators.

The Future of Money
Bernard Lietaer 2001

Wörgl

money that is not worth keeping

Have you ever wondered at the beauty of buildings from the 12th and 13th centuries in southern Europe, when about 600 new towns were built with solid stone houses that have survived to this day? Ironically, it was not a politically stable period. The wars waged by the Church on the Cathars necessitated much reconstruction. The wars were partly about religion and partly about the Church's loss of revenue.

Inadvertently a rather strange system of money was responsible for the reconstruction that gives us such pleasure today.

In the Holy Roman Empire rulers of the dozens of small independent states issued thin silver coins called *bracteates* which would be reclaimed at three quarters of their value when the ruler died. Realizing that this was rather a good idea the rulers soon got the idea of reclaiming and reissuing them more frequently. So holding on to money was risky – it was no store of value; **better to use it while you had it to build houses with lasting value.** This led to a high demand for construction labor so wages were good, working hours were short (about six hours a day) and there were at least 90 religious holidays a year. It greatly improved the quality of life and was wonderful for the local economy.

But when gold coins were introduced in the fifteenth century it became worth storing money instead of using it because of the amount of precious metal it contained. You could keep it under your mattress. So demand for labor dropped, wages fell, unemployment appeared, some businesses closed down because people could earn more by lending the money they had accumulated on interest rather than by trading. And, to cap it all, rulers had to find other means of taxation.

The idea of issuing money that is not worth storing has cropped up frequently since. Wörgl, a small town in Austria, was suffering from the Depression in the 1930s. There was 35% unemployment and local taxes were seriously in arrears, crippling the work of the council. The mayor negotiated a loan from a credit union bank and printed scrip notes. These were guaranteed against the national currency though you only got 98% of their value if you exchanged them. You had to stamp each note monthly to one percent of its value, so it cost you to keep it and you used it as quickly as possible. Half the council's staff wages were paid in the scrip and the mayor also allowed local taxes to be paid in scrip, so businesses and shops were prepared to accept it. The effect was to stimulate local trade, tax arrears were paid off and there was full employment. The council was able to employ 50 extra people, paying their wages entirely in scrip, to repair and surface streets, and to extend the sewerage system. Then they built a ski jump and a reservoir without incurring any debt. The scrip notes circulated around the local economy so there was little call on the inflating national currency. Not surprisingly, other towns started to copy the scheme. The Central Bank, alarmed that it was losing control of the money system, took out legal proceedings and closed it down soon after it started.

Many similar schemes were started in the United States with the same response from banks. In March 1933 President Roosevelt forbade any further scrip issues being advised that the financial system was being taken out of the government's hands – by the people!

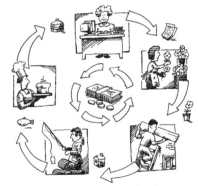

economic activity using a local currency

Short Circuit
Richard Douthwaite 1996

Global Eco-Currency

four currencies

Anyone who takes part in a baby-sitting circle uses an alternative currency, it may be IOU chits, entries in a diary or just a mental note; there is nothing odd about multiple currencies. It is very odd, however, to use just one currency for speculating on the international money market, buying cabbages, and saving for your old age. Richard Douthwaite suggests four currency systems each designed for a specific purpose:

- Ⓢ an *international currency* for trading between nations;
- Ⓢ a *national-exchange currency* for trading within a nation,
- Ⓢ various *user-controlled currencies*, and
- Ⓢ a *store-of-value currency* for your savings.

The *international currency*, would be an emissions-backed-currency-unit, an 'ebcu', and take its value from the right to emit greenhouse gases (see Contraction & Convergence). There is no international currency at present: various national ones such as the US dollar, the British pound, the yen and the Euro are used instead. **As these are debt-based and not linked to anything real, the global economy is extremely unstable.**

To get the new currency going, the IMF would issue each country with 'ebcus' based on its 1990 population. No more would ever be issued. A nation's existing wealth would remain in its other currencies. Then, each month, the IMF would issue coupons called Special Emission Rights (SERs) to each country on the same population basis. These coupons would confer the right to emit greenhouse gas and would be handed in whenever fossil fuel was purchased. Each year, the number of SERs issued to each country would fall so that, within a few decades, the total amount of greenhouse gas the world released would match the earth's capacity to absorb it.

National governments would decide how many of their SER coupons to distribute internally to enable their citizens to use fossil energy and how many to trade internationally. India, which uses little energy at present, would have surplus SERs to sell. The US would use all its coupons internally and need to purchase more on the international market using its 'ebcus' in payment.

The *national-exchange currency* would be used for buying and selling within a country but not as a store for one's long-term savings. If trade increased the central bank would print more of this currency and spend it into circulation, thus reducing the government's tax needs. This might cause mild inflation but this is fine because it induces you to use money before it loses value, thus encouraging trade and employment.

The national-exchange currency would be supplemented by a wide range of *user-controlled currencies* such as Local Exchange Trading Schemes (LETS), scrip issues, Time Dollars and systems modeled on the Swiss Wirtschaftsring.

The *store-of-value currency* would be a fixed quantity of a special form of money used only for purchasing and selling capital assets. If more houses or shares were traded, for example, the need for this currency, and hence its value, would rise in relation to the national-exchange currency. So:

⑤ The *international-currency* would keep the world economy within the carrying capacity of the planet.

⑤ The *national-exchange currency* and many *user-controlled currencies* would encourage economic activity within a nation.

⑤ The *store-of-value currency* would link your savings to the prosperity of the country – no more and no less.

Other thinkers suggest incremental change. But, with the likelihood of a global crash, it might be better to adopt Douthwaite's radical proposals than to try and patch up our present perverse financial system.

The Ecology of Money
Richard Douthwaite 1999

Citizens' Dividend

for a civilized civilization

> "The time has come for us to civilize ourselves by the total, direct and immediate abolition of poverty."
> *Martin Luther King Jr.*

The United States is the wealthiest nation the world has ever seen but who created this wealth? It was built up from the time of the Declaration of Independence and before by the government, by the railroads, by the roads, by the farmers, by the imagination and work of countless individuals over many generations. This is our inheritance.

But the horrific inequalities of corporate capitalism have resulted in 220 people owning more than the joint wealth of half the world's population. The wealth of these people doubled in just four years. Soon the entire planet, humanity's inheritance built up over centuries, will become the private property of just a few individuals. Is this the New World the American Founders struggled to create?

A new concept is emerging, that the modern world's inherited wealth can eliminate poverty. The idea started in Europe where the concept was called the Citizens' Income. In New Zealand it is called the Universal Income. In South Africa it is the Basic Income Grant. In Canada the Guaranteed Annual Income. In the US it is the Citizens' Dividend.

A Citizens' Dividend would be paid to every woman and man in the country, equally. It is not welfare or charity. It is their rightful portion of the nation's wealth. Attempts to alleviate poverty until now have all been indirect – the welfare program, housing, education – why not be direct? Abolish poverty.

The change would be dramatic. The Dividend would give income to millions that at present have no purchasing power. There would be an explosion of local economic activity, and the crippling stress of modern life would be removed.

A change of this scale is difficult to imagine. But could anyone have seriously imagined, or wanted, the world we are in the process of creating? Where 220 people own the planet? Where the rest live in fear of loosing a job, of not being able to pay a mortgage, of falling ill, of penury in old age? The fear of starvation shared by a billion people in the world? A fear that drives people to vicious competition or theft to gain elusive security? The fear that drives businesses to expand regardless of the necessary sweatshops or their destruction of the environment, just in order to survive?

A better world is possible. And leading economists have said that this kind of society can be afforded. J. K. Galbraith calculated that a Guaranteed Income (least change from current procedures) would cost the US only $20 billion a year. Remember, the US spends $400 billion a year on armaments.

No worries!

A more radical approach would draw the money from funds that are not part of the market economy; from government activities, the monopolies, the government-granted privileges, inheritance, rent, interest and special breaks. It would come from all those things that are national "commons." For example, the airwaves that are licensed to private broadcasters. Alaska already distributes royalties from oil drilling companies to all its citizens, equally; this is only about $1,200 a year, $4,800 for a family of four, but it has boosted the local economy and it demonstrates the principle of a Citizens' Dividend. The amount of the Dividend would not be an arbitrary figure decided by the current administration. It would be an equal portion of the unearned wealth of the country. A preliminary calculation indicates that this might be around $11,000 for each citizen each year, though Robert R Schutz suggests it could be as high as $30,000.

A Citizens' Dividend would free people to improve their lives. It would provide them with a financial platform from which they can choose the life they want to lead. They could get out of work they find unsatisfying making way for others that want the job. They can avoid work that harms the environment. Small-scale farmers could intensify their holdings. Local shops within a community would again become viable. The trend that at present centralizes all economic activity would be reversed, making it possible to curb the tyranny of corporations and agribusiness. Employers would benefit because of lower staff costs, since they would only need to top up the Dividend and could therefore employ more staff, spreading the workload. There would be no need to produce unnecessary or short-lived products just in order to generate employment. And people would not need to amass excessive wealth for their old age.

Some people might lie around and do nothing, though there would be no justification for them to scrounge or beg. But most would want to earn more than the Dividend. Some might choose to bring up a child or write a book while being content with a low income.

A Citizen's Income would do more than anything else to give us a humane and civilized society.

▶ The BIG idea

Unemployment among South Africans runs at almost 50%, making means-tested welfare payments, if attempted, too expensive to administer with any pretense at fairness, and would encourage corruption. The World Bank and IMF forced the government to impose "structural adjustments" that deprived the poor of health and social provision but did nothing to stimulate the economy.

Something new is at last giving hope. The South African government is considering the BIG idea, the Basic Income Grant. This grant would be given to all citizens and it would be clawed back through tax from people with an income. A BIG coalition has been formed of labor unions, church bodies, NGOs and organizations like the SA New Economics Foundation. All eyes are now turned on South Africa to pioneer this new people-centered economic idea.

The $30,000 Solution
Robert R Schutz 1996

Vibrant Cities

creators of prosperity

Back in 1950 Taiwan had a rural economy with low wages that attracted companies from America. These factories provided some employment but did little to stimulate a local economy. Taiwan might have remained a backward supply region like so many poor countries today.

It was not global free markets or aid that changed things; it was its government. It distributed land to peasants and, crucially, attached strings to landlords: they had to invest part of their compensation in local light industry. The government also used training grants, tariffs, quotas and export subsidies to protect the emerging industries. People learned skills from the US firms, they were sub-contractors, they replaced imports with locally made goods, they invented new products and they financed their new businesses with the local money from former landowners. The city of Taipei and its region became a hive of

activity operating first locally, then trading in the Pacific Rim with cities at the same state of development, then globally.

A "supply region," whether it produces minerals, food or oil, benefits only as long as someone is prepared to pay good money for its produce. During this period it may be wealthy enough to import all the gadgets it needs for a diverse and prosperous lifestyle and does not need to make these things itself. But, when other countries no longer need its goods or when they can drive the price down, it can no longer afford imports and cannot make them itself. It slips back into poverty. However if, during the prosperous period, it replaces imports with locally-made goods and focuses its economic activity in cities, it can build an economy that retains prosperity.

What has produced strong city regions in the past? Ports had a natural advantage when ships were the main international transport. Cities near coalmines prospered until oil and electricity arrived. Some cities, like Seattle, still benefit from military contracts. European cities developed economic strength because each had its own currency that would fluctuate to encourage exports or allow imports depending on the business cycle. Tariffs enabled America to break its dependence on Europe.

Tariffs, import quotas and protection against takeover by foreign investment are essential tools for poor countries – much better than aid. A network of city regions at a similar stage of development, with fluctuating currencies, is the best environment for developing their trade. Above all, they must avoid simply selling basic commodities for others to manufacture, or agricultural produce for others to process. These regions are referred to as "import replacement economies" because their vitality arises from replacing imports with increasingly sophisticated local products. Rural areas need to be part of one of these vital regions in order to get adequate value for their more basic produce.

In the West, cities no longer have their own currencies; it is now national currencies that fluctuate. These fluctuations are determined by the needs of a dominant city region, so London flourishes while the north of Britain declines. The Italian currency suited Milan but not the south. Each Scandinavian country, however small, retains one viable city region, but only one. The Netherlands, sensibly, has grouped its cities into the Ring City as a single entity. Singapore, Taipei and Hong Kong each has its own currency, which partly explains the spectacular success of the Pacific Rim. The single European currency will probably result in just one city region becoming dominant while outlying areas decline - unemployment in Portugal and East Germany is already rising. America is a special case because the inflow of goods in exchange for the use of the dollar across the world has prevented the trade deficit detracting from its internal prosperity.

Indians have a genius for improvisation and import replacement. The country's future may depend on being able to resist the pressures of a global economy that aims to make its rural areas, once again, simply supply regions for a western empire. India has a sixth of the world's population with a market large enough for a full diversity of internal trading. Trade with Europe and the US should be limited to high value goods where the value added has been made within its own borders. India's great city regions are widely separated; each could adopt a separate currency and, by trading with each other using fluctuating exchange rates, develop self-reliance and sophistication. Like the Pacific Rim.

The single global patent regime prevents the spread of knowledge and hence the ability of poor countries to add value. India's 1970 law rejecting certain western patents enabled it to replace imports with local production; and its cheap generic medicines have saved millions. It was only worldwide protest that allowed South Africa to use affordable drugs from India to treat AIDS (a recent agreement will once again prevent this). The corporations fear loss of income for research but their R&D budget is tiny compared with the money they spend on marketing. And only a tenth of the R&D budget goes on drugs relating to 90% of global disease, the bulk being spent on rich-country afflictions like obesity.

 Cities and the Wealth of Nations
Jane Jacobs 1984

Mosques & Churches

misunderstandings abound

A group of building workers line up. One stands in front to lead the prayers. Draw a line around this group, put a bulge at the center of the side facing Makkah and you have the plan of a mosque.

From the noisy, chaotic activity of a bustling covered *suq* in Damascus, they pass through a small door into a large calm collonaded court where groups of men and women sit on carpets in the shade and talk, read or sleep. Mosques are enclosures for the community and fill for prayers five times a day. Churches have different qualities. In gothic cathedrals holy rites are performed at the altar and the architecture soars towards heaven. Orthodox churches embody mysteries shrouded in incense. Protestant churches, chapels and assemblies encourage the congregation to focus on a preacher.

Muhammad had been both prophet and head of the community and there was no separation between religion and state. The *Qur'an*, revealed to him over a period of twenty-three years, insists that a Muslim's first duty is to create a just, egalitarian society, where poor and vulnerable people should be treated with respect.

There are five "pillars of Islam." First, a Muslim is required to make a brief declaration of the unity of God. The other four emphasize social equity and community: All should pray five times a day; this is a communal act that is done, not just thought or said. *Zakat* almsgiving should ensure fair distribution of wealth in the community. Many Muslims see western values of competition and self-interest as being a direct threat to the kind of society to which they aspire. The *Ramadan* fast is a vivid reminder of privations suffered by the poor. For the fifth pillar, *al-hajj*, pilgrims wear a uniform white garment that allows no difference of rank or standing. With its care for the community, Islam appeals to the poor and is the fastest growing religion in the world. The Muslim is a social being in the body of Islam, the *ummah*, that transcends nationality. When a Muslim nation is ravaged, the hurt and shame is felt universally and by each individually.

Muhammad did not believe that he was creating a new religion but was bringing the primordial religion of humanity to his Arabian tribe. God had sent messengers to every people on the earth to tell them how to live. When they disregarded these basic laws, creating tyrannical societies that oppress the weak and refuse to share their wealth fairly, their civilizations have collapsed. This required *jihad* ("struggle") for right living. Seventh century Arabia was a country of warring clans which may explain some harsh phrases in the *Qur'an*, but they are always followed by injunctions like "if they let you be, and do not make war on you, and offer you peace, God does not allow you to harm them." War is only permissible in self-defense and is always evil.

Southern Spain was, for hundreds of years, a Muslim country that preserved civilization in Europe and embraced thriving Christian and Jewish communities. In 1492, Ferdinand and Isabella made their first conquest. Muslims and Jews were forcibly converted, massacred or expelled and many Jews fled to the Ottoman Empire where they were welcomed. Western histories gloss over the barbarities of the Christian invasions and crusades. But the tables turned; Western modernism eventually won and the Ottoman Empire collapsed.

Just as the United States' weapons of mass destruction do not sit easily with the Sermon on the Mount, Osama bin Laden's terrorism does not sit easily with Islam. But the social and egalitarian aspects of Islam's five pillars may help to explain why it appeals to the poor and why it is the fastest growing religion in the world.

 In Santarém, Portugal, the mosque, after 700 years of disuse, is being restored along with the synagogue and the monastery of Sao Francisco. Might they again experience the intellectual and cultural dynamism that was once there?

The Battle for God
Karen Armstrong 2001

Basic Needs

isn't everyone as happy as we are?

Helena Norberg-Hodge worked in Ladakh, in the Himalayas, one of the harshest climates in the world, scorched by the sun in summer and frozen for eight months of the year. She noted the immense care taken over the use of minimal resources, relying on the outside world only for salt and tea – there was literally no waste. Ladakhis worked hard in the summer but at their own rate, accompanied by laughter and song; the distinction between work and play was fluid. Winter was a time for festivals, parties and story-telling. They understood their own needs - and these did not relate to western style development.

After visiting this poorest of countries for 16 years she said, "Their sense of joy seems so firmly anchored within them that circumstances cannot change it. At first I couldn't believe that the Ladakhis could be as happy as they appeared. It took a long time to accept that the smiles I saw were real." Is this "noble savage" romanticism or do we need to be more precise in defining human needs?

Manfred Max-Neef, a specialist in human-scale development, identifies nine basic needs:

- 😊 **Subsistence –** *Creation, health, food, shelter skills, work, feedback*
- 😊 **Protection –** *security, society*
- 😊 **Affection –** *friendship, family, love*
- 😊 **Understanding –** *curiosity, education*
- 😊 **Participation –** *responsibilities, interaction, community*
- 😊 **Leisure –** *play, fantasy, intimacy, privacy*
- 😊 **Creation –** *skills, work, feedback*
- 😊 **Identity –** *belonging, groups, recognition*
- 😊 **Freedom –** *autonomy, rights, dissent*

But basic human needs can be satisfied in very different ways by different cultures. If we propose to make changes to the economic system of a country we should check how the changes are likely to affect any one of these nine basic needs. To facilitate this he discusses each need against four aspects of living – what we are – what we have - what we do – how we interact.

Severe deprivation in one need may paralyze all other considerations. This is particularly the case with hunger (subsistence) that paralyses 13% of the world population. But the same is true of other needs. Social exclusion in an affluent society may lead to criminal behavior. Total lack of affection or loss of identity may lead even wealthy people to suicide.

For politicians and bureaucrats the concept of basic human needs is usually limited to food, health and education – to be met through aid and advice handed down from international agencies. They see only the economic goods that represent a fraction of real need and are surprised when imposing these solutions, with the best intentions, causes havoc in countries with a strong spiritual and traditional culture.

For Max-Neef human-scale development comes from the grass roots. An economic strategy may satisfy one need and not another. If it causes extreme deprivation to any one of the needs it must be abandoned however beneficial it appears to be in other respects. Development will be destructive unless those involved participate fully in deciding what their real needs are and how they might be satisfied.

Back to Ladakh: globalization may bring additional income but the influx of money and consumer perishables may simply make them aware of the deprivations of an extreme climate while destroying a culture that in the past has satisfied all their fundamental human needs.

In 1995 it rained in Ladakh for the first time in living memory. Their houses are built for a dry climate. Climate change? They have done nothing to deserve this.

 From the Outside Looking In
Max-Neef 1992

"Them" or "Us"

a Gandhian approach

Henry VIII is usually credited with creating Britain's centralized state, a system that Thomas Jefferson later struggled to reject. Everything there now flows from central government. Gandhi also suggested precisely the opposite. Various projects in India are applying his principle: that **the local community, at grass-roots level, should be the basic unit of government.**

Seventy-five villages at Lathur, in South India, have been identified as being below the poverty line and in need of aid. But it is the village (100 to 300 people), not the authority, which decides which individuals are in need and how the aid should be distributed. The village council of 20 men and 20 women is elected annually and the post of headperson rotates.

Mr Rajamani could double his income if his well were deeper. He applied to the village council. The village council assessed his proposal and approved the necessary loan (from the authority). He repaid the loan plus interest to the village council, not to the authority. This is the route by which aid flows to the village community.

Ninety-eight percent of loans are satisfactorily repaid and the failures are usually for a particular reason that can be understood by the village council. So the community builds up a fund. Once the fund reaches a certain size the authority withdraws and the village itself continues with the process using its own accumulated funds. After five years, 35 of the villages no longer require further aid. A community bank, controlled by the village councils, handles the funds.

It is necessary, of course, to ensure that each village keeps to the system and is not bullied by a few individuals or taken over by a political group. This is done not by the authority, but at an annual two-day gathering of all the villages. In addition to the business in hand this gathering becomes a festival, a place of contact and even a marriage market. It is the lowest tier of the state electoral system that Gandhi recommended, where each tier takes responsibility for its own affairs and aims to determine the policy of higher tiers of authority. The higher authority must be the servant of the lower. In other words, the villager is the maker of his/her own destiny, and is his/her own legislator, through his/her chosen representative. The ultimate storehouse of his/her power, should other means fail, is non-violent civil disobedience.

In Britain, even more than in the US, power is now held by a clique at the top subject only to occasional elections – a system they call democracy, though a political analyst would call it "elective oligarch," rule by a small group of people. Each tier distrusts the one below, so rules and regulations abound. The lowest tier controls thousands. Local authorities are reluctant to allow a street community to decide how its road might be allocated between children and cars, or to approve the design of new buildings; to set up its own arrangements for repairs, or re-cycling of waste and water; or to combine with adjacent communities to provide its own school or clinic. A street community in Hull, England, secretly grassed its road overnight for the Jubilee in 1977 and had it ripped up by the authority the following week. It took 20 years for the community to get it back to grass through official channels. Citizens only really feel involved when they themselves make the decisions that vitally affect them. There is a fundamental difference between decisions made by "US" and decisions made by "THEM."

It is no wonder that social exclusion has become a major issue in Britain. When people cannot determine their own destiny, the community falls apart. The vital step is from mere consultation to the exercise of citizen power through participation. This requires the management of budgets by communities that are small enough for all to be involved – street-scale communities. This is democracy.

Reclaim the Streets

Two old cars collided. The drivers shouted, took axes from their trunks and hacked the other's car. Traffic gridlocked. Onlookers cheered. That's how the first street party started in 1995. On another occasion 3,000 danced to the tune of "What a Wonderful World". Trees were planted in a six-lane highway under the hoop skirts of clowns on scaffolding. The largest party drew 20,000 in central London. Then it went international with a Global Street Party in 1998 on the day that G8 leaders met. In Prague police action provoked violence. In Sydney, Australia, police helped and joined in the fun.

Email helped. People converged, as if from nowhere. The number of bicycles suddenly multiplied, "We're not blocking traffic," the Critical Mass riders said, "we are the traffic." People made their own fun without asking the authority's permission or relying on any corporate largesse. The urge for entertainment, through acts of civil disobedience, became festivals. The streets, for a brief day, become commons.

The realization had dawned that our streets had been colonized by the authorities, by cars and by commerce – just as France had been colonized during WWII. It was time to tell the authorities and big business, the streets belong to us not to you.

"Freedom is to be attained by educating the people to a sense of their capacity to regulate and control authority."
Mohandas Gandhi

Fences and Windows
Naomi Klein 2002

Wealth in Poverty

a tribal's view

Wealth means different things to different people. For a tribal person, money has little meaning. It is community that they value.

The community at Gudalur, South India, is extremely poor. They used to live in the forest but could not prove ownership of any land. In their culture there was no conception that land *could* be owned. **Land, water and air are regarded as commons, available for all to use.**

The government sold the forests in which the tribal community *adivasis* lived. It was assumed that the forests were empty and Brooke Bond acquired large tracts as tea plantations. Tribals continued to live on the edges. But how could they secure their rights? The only way was to plant permanent crops. But when they did so

the foresters pulled up the crops and destroyed their homes.

In 1984 Stan and Mari Thekaekara, social activists, persuaded five of these tribes, totaling 2,000 families, to act together. When individuals were threatened, the community gathered to protest. They secured land for some, but ownership of individual plots did not help the community.

The *adivasis* decided that the only way to strengthen the community was to obtain a tea plantation. They prepared a business plan and, with Stan's help, obtained a loan to buy it. The plantation is now their own. In two years the yield doubled, and it supports a school and a hospital. From being virtual outcasts, the *adivasis* are now respected. Their tea fetches high prices because of its quality, but; rather than dividing the profits among the group working on the plantation, the pickers themselves said that any extra earnings should go to the community. **They said that if they became wealthy while others remained poor, their community would fall apart.**

Mari said: "The Adivasis were clear about their wealth. It is our community, our children, our unity, our culture and the forest. Money was not mentioned. We, the non-adivasis, were stunned. As we discussed concepts of poverty further, we realized that they didn't see themselves as poor. They saw themselves as people without money."

Stan and Mari visited the poor Easterhouse estate in Glasgow, UK, as part of a North-South exchange and the question of poverty again forced itself upon them. In Glasgow everyone had a television, running water, heating and benefit payments. The *adivasis* would regard these things as inconceivable luxuries. "But most of the men in Easterhouse hadn't had a job in 20 years," Mari said, "They were dispirited, depressed, often alcoholic. Their self-esteem had gone. Emotionally and mentally they were far worse off than the poor at Gudalur, though the physical trappings of poverty were less stark."

Stan and Mari brought a group of *adivasis* to Germany. The gift they valued most from their hosts was being treated as equals – something they did not experience at home. But they were speechless when they saw an old people's home. "How can children send their old parents to live alone? We must ensure that such things never happen in our society, no matter how much we progress." Then Karl, their German host, came home one evening preoccupied with the news that he might lose his job. Bomman, an *adivasi*, worried all night for his new German friend. "I have an idea," he announced in the morning, "I can make bamboo flutes when I go home and Karl can sell them here till he finds a job." Through contacts like these the *adivasis* now trade directly with their friends in Europe.

When a group was next threatened with eviction, it was the contact with their friends in Germany that persuaded Brooke Bond to concede – confrontation might damage their image internationally. The *adivasis* had found that community, with its ability to empower, even when up against a trans-national company, could be extended across the world.

Is this just a story about a remote community or does it point to human values deeper and more lasting than those of global capitalism?

🐟 Free trade can hurt. A letter received from Ramdas in Gudaur in July 2000 said, "One of the most unfortunate issues has been the drop in price of both tea and coffee. The government has opened up the Indian market and large imports are being allowed. This has created a fall in the price for producers while the price in shops remains the same."

🐟 Kariyan, of the Kattunayakan tribe in south India, made a clearing in the jungle and planted ginger. As luck would have it, the price for ginger shot up last year from Rs60 to Rs160 and he made a lot of money. Now he has none left. When people asked, he gave, without asking questions. Does he need lessons in money management?

Kariyan values the tribe. He shares what he has with others and vice-versa. He did not calculate, he acted as he and they had always acted. If he had gone off on his own, the money would have eventually run out and he would be destitute. He has security as long as the tribe has security, which means as long as they all act instinctively as he did. Kattunayakanis do not need to store money for their old age, they do not consume more than they need, they do not cause harm to the forest or its animals because they have a strong sense of interdependence with their surroundings. Instinctively. The Kattunayakan tribe has facial features resembling Australian aborigines, and anthropologists confirm the connection. The startling deduction is that the tribe has had continuity from the time when humans migrated to Australia 50,000 years ago. They have a sustainable way of living.

"There is no quiet place in the white man's cities. What is there to life if a man cannot hear the lonely cry of the whippoorwill or the arguments of the frogs around the pond at night?"
Chief Seattle

Just Change

beyond fair trade

A community of latrine workers in Gujarat, disparagingly referred to as *bhangis*, drink a lot of tea. A tea plantation owned by *adivasis* (tribals) sent them a truck load of organic tea. The *bhangis* drank some themselves and sold the rest to teashops at 80% of wholesale price. These shops would not normally buy anything from *bhangis* but at 20% discount the barriers of caste crumble.

It is a co-operative and no trader is involved. The pickers get their normal wage, the truck driver is paid his costs and time and the *bhangis* receive something for their time in distributing the tea but deduct the cost of the tea they drink. The tea that is sold to teashops covers these costs and produces a profit. But who should benefit? The *bhangis* met the *adivasis* and it was agreed that the division of profit should be according to need. In this instance the *adivasis* were so horrified by the conditions forced on the latrine workers that they insisted the *bhangis* kept all the profit.

Stan and Mari Thekaekara, social activists working with the *adivasis*, are in touch with some housing estates in Germany and England that drink a lot of tea and coffee. Sample blends were sent so that they could choose the tea or coffee they liked. A load was delivered, more than the estate needed for itself, so it set up in business to distribute the tea at local events and in local shops. Exchange visits put the two communities in touch with each other and email helps to keep these friendships alive. This has produced benefits for all involved – the person who advanced money for the venture was reimbursed at cost plus a negotiated share of profit, the *adivasi* pickers received a guaranteed wage and people on the housing estates gained employment.

Apart from housing estates, there are many other organizations like schools and local authorities that already have an information and distribution network and do not need advertising or smart packaging when choosing their products. These organization's have ethical concerns and would welcome the chance both to make links with communities in poor countries and to avoid dependence on the commercial world. At the Indian end lentil growers, textile workers and fishing communities are joining in.

Just Change, as the process is called, aims to take forward the excellent objectives of the fair trade movement of companies like Cafédirect or direct traders like Bishopston Trading, by involving communities with each other. It is a new model for trade between communities that can call each other friends.

➤ An earthquake struck Gujarat in January 2001 causing the *banghis* to lose houses and possessions. The *adivasis* immediately donated a truck load of clothes and supplies to their new friends. Krishna and Yogananda, architect and engineer, went to Gujarat to see which of the traditional houses were unaffected by earthquake. They urged the Gujaratis to use their plentiful labor to build cheap permanent one-room extendable structures rather than calling for temporary tents from aid agencies. Bamboo, with its long strands and nobles, is good for reinforcing mud-cement walls against earthquakes but the Gujaratis were not familiar with it. So a group, whom Krishna had trained in Gudalur, volunteered to come and teach them techniques used in the south. Just Change demonstrated its value in an unexpected way.

Odious Debt

a humane American idea

It appears that international law has one rule for the strong and another for the weak.

The concept known as "odious debt" was established by the US in 1898. America had invaded Cuba but found that their new colony had large debts to Spanish banks. The US refused to pay these debts, arguing that the debts had been "imposed upon the people of Cuba without their consent and by force of arms." Thus the doctrine that neither the people of a country nor its new government is responsible for the odious debts of a previous regime was established. However, do you imagine that in recent years poor nations were allowed to invoke this doctrine? Certainly not!

South African apartheid was defined by the United Nations as a crime against humanity. When Nelson Mandela became president, he and the people of South Africa inherited more than $18 billion in debts. Surely no debts could be described as more odious. But the IMF warned that unless the debts were repaid, however odious, South Africa would be isolated by the international community. Money that should have been used to build schools and homes, to create jobs and to repair the environment, instead was sent to the US, UK and Swiss banks that had financed apartheid.

Mobuto, dictator of Zaire, was a friend of the North and the North lent him money. He was known as the "kleptocrat" because of his lavish palaces and purchases from Europe. When he died in 1998 he left debts of $13 billion. Zaire is now the Democratic Republic of Congo and the West wants its money back. The new government should be spending every penny on rebuilding the shattered country. Instead, every man, woman and child in the country must repay $260 in debt.

Investors lent the Philippine's Marcos regime the money to build the Bataan nuclear power station, and Westinghouse channeled millions of dollars through Swiss banks to secure the contract. It was in an earthquake zone so was never used. The Marcos' personal wealth was $10 billion, but it is the Philippine government that must pay the debt.

Poor countries must pay their debts, however odious. But Long Term Capital Management, a US-based hedge fund, was treated with concerned sympathy when it landed itself in debt. The Federal Reserve immediately found $3.6 billion to bail out its friend.

Dealing with the Asian crisis, the IMF declared, "Reducing expectation of bailouts must be the first step in restructuring Asia's financial markets." Asian companies were bankrupted and became rich pickings for western companies. However the IMF ultimately agreed $120 billion funds to be made available to pay Asian debts to – whom? To western banks, of course. Thus the West not only took over Asian businesses at knock down rates, it also got its money back.

The guiding principle for the IMF and the World Bank is this: whatever happens, whoever is at fault, the wealth of western creditors must be protected and enhanced.

Value of life

World organizations have a primitive creed: dollar, euro and yen. Someone with a billion dollars has great value, someone with ten thousand not much. Someone without money has no value; he or she is outside the market and therefore outside the civilized world. This religion demands human sacrifice and peasants in poor countries, taken from their land, provide plentiful candidates.

Third World Debt

view of the rich

"The notion that for the Jubileum someone can come along and forgive that debt is whimsical. If you have a society based on debt forgiveness, who's going to invest in debt anymore? So you really screw up the market."
Mr. Wolfensohn, World Bank President

Poor nations owe the rich a lot of money, $2,500 billion. Of this, $200 billion is owed by 41 Heavily Indebted Poor Countries (HIPCs). But the HIPC debt is small compared with the total.

The First World, say the rich, gave aid and loans in order to help the Third World develop its economies and to combat poverty and disease. In addition the World Bank has been providing health, education and agricultural advisers. The Agrochemical Revolution, pioneered by First World scientists, greatly increased productivity since the 1950s and helped poor countries adopt modern techniques to replace traditional farming methods. Much of the aid was squandered by corrupt regimes and many countries mishandled their economies. With this perspective, the First World argues that it has made serious efforts to help poor countries rise out of poverty.

But northern governments and the World Bank can be accused of financial malpractice. If a bank lends to a corrupt or chaotic company that then goes bust the bank loses its money and must write off the debt. But the First World knowingly lends to countries that cannot possibly pay back the capital let alone the interest. It also lends to corrupt regimes for political reasons. The debts are calculated in hard currency, usually dollars, so they increase in real terms as the local currency devalues. On top of this the debts attract compound interest. But the North never writes off the bad debts – they accumulate and multiply. **So the total Third World debt is three times the amount actually lent.**

The money owed by HIPCs to commercial banks was miraculously converted by the G7 (group of seven rich nations) into debt owed to public institutions like the International Monetary Fund (IMF). And through them to taxpayers. You and I must now pick up the tab for the malpractice of commercial banks. The cynicism and immorality of this is staggering – the G7 was prepared to release commercial banks from their debt but not the countries that had to divert money from health, education and social provision.

But why don't poor countries just refuse to repay unfair debts?

Nicaragua's health budget is a quarter of the amount it spends on servicing debt. Mali repays more than it spends on health and one in four of its children die before the age of five. Zambia's repayments are greater than its spending on health and education. Debt repayments cripple struggling economies and cause hunger, disease and death. If a dictator imposed the policies of the IMF he would be found guilty of genocide.

By contrast, the financial assets of northern banks grow at the rate of $2,500 billion a year – the amount of the entire third world debt. As for aid – debt repayments by the Third World are nine times as much as the aid it receives.

At a meeting in Cologne in June 1999 the G8 made a gesture to the HIPCs, reducing their debt by $100bn. Campaigners were jubilant and there was dancing in the streets.

They had forgotten that the IMF and the World Bank were involved. These require structural adjustments before any money is released – measures like opening markets to free trade and allowing global corporations to take over essential services. Four years later $29bn had been cancelled and $24bn added making them only $5bn better off (remember the US spends $400bn annually on armaments). Tears are shed from time to time for the plight of desperately poor African countries but that does not stop us looting the continent's resources.

Third World economies are largely dependent on selling commodities, e.g. minerals and crops. But WTO policies are pushing the prices of commodities to an all time low. IMF structural adjustments force poor countries to reduce social spending. In the few instances where debt has been cancelled, spending on health and education has more than doubled.

The First World however, has put its head into a noose. If a man owes a bank $10,000 the bank has him in its power. If he owes it $10 billion he has the bank in his power. **The Third World owes $2,500 billion and has the global economy in its power.** At the WTO meeting in Cancun in September 2003 the majority nations had their first experience of the power behind collective action. They can either bring the global economy crashing down, noting that after 55 years it has failed to produce an equitable world, so why persist? Or they can negotiate gradual reduction of debt and a new world order based on environmental survival and social justice.

- Poor countries are losing $700 billion annually in earnings due to the trade barriers of industrial countries, says Renato Ruggiero, former director of the WTO.

- Annual expenditure on health per capita:
  US $2,765
  Tanzania $4

- "IMF rescue packages are intended only to rescue western creditors."
  Ellen Frank, Professor of Economics, Boston

- Ecuador's debt repayments amount to nearly a third of her budget, twice the budget for social welfare. Nearly two-thirds of Ecuador's population live in poverty.

- In July 2002 over three million people faced starvation in Malawi. The World Bank and IMF had forced Malawi to sell 28,000 tons of maize to repay dollar debts.

- Between 1503 and 1660, 185,000 kilos of gold and 16 million kilos of silver were transferred from Central America to Europe. One would hesitate to accuse Europe of looting or stealing; it was surely a loan. Even if Central American nations only ask for modest interest, perhaps half of what we now charge the poor Third World, the debt we now owe them would have 300 digits.

"Only when the last tree has died and
the last river been poisoned
and the last fish been caught will
we realize that we cannot eat money."
a First Nation American

First World Debt

view of the poor

Money is not the only form of wealth. Would you envy a billionaire his money if he were locked in a prison cell? If we can't enjoy the world around us we are indeed poor. So there are two kinds of wealth: your money and your environment. There are also two kinds of debt – the money you owe and the damage you have done to the environment.

The First World says it is owed lots of money. And not in devalued rupee, baht or rouble but in hard western dollars, so the original debt may now be multiplied by five or 10 times due to devaluation and compound interest.

The Third World has a very different view about debt. They are saying the word "third" is an insult from people with myopic minds set on money. We have qualities of culture, spirituality, heritage, traditions, festivals, cuisine and language quite as good as yours (and our children are not obese).

Who has been using up the earth's fossil fuel, mineral and timber resources?

Who has been polluting the atmosphere with greenhouse gases and causing hurricanes and rising sea levels to devastate our countries?

Who has been profiting from the chemicals that have got into our food chain, degraded our soil and poisoned our aquifers?

Who has been over-fishing the oceans and depriving us of our offshore fisheries?

How dare the Money-First World require us to make "structural adjustments" to our economies when it makes no structural adjustment to the social harm and environmental damage it is causing throughout the world?

They say that the First World has a massive debt to other nations for depleting the natural capital of the world, but this debt is fundamentally different from Third World Debt. Financial capital claimed by the First World is artificial; it is a human invention that is meant to conform to rules largely invented by the Money-First World. Natural capital, on the other hand, is real and finite. It is the fossil resource fixed in the earth's crust through millions of years of photosynthesis; it is the constant daily flow of energy from the sun; it is the ability of the atmosphere to keep the earth pleasant and habitable.

The majority poor nations could have a case in law against the minority rich. If pollution by industrialized nations has led to climate instability that is causing drought and famine in the Horn of Africa, sinking island states, causing flooding in Venezuela, Bangladesh, Orissa and Mozambique – the industrialized nations should pay compensation. Attempts to quantify this liability have an air of fantasy – you can quantify loss of infrastructure but how do you quantify loss of life, loss of livelihood, loss of culture? The debt is infinitely greater than the whole of the Third World financial debt.

The UN World Disasters Report 2000 calculated that the First World had amassed a debt of $13 trillion to the Third World. Add to this the poverty and deaths caused by a perverse global economy and unfair trade practices. Add to this the deprivation and deaths caused by its environmental damage. The Money-First World has been responsible for a sustained holocaust in the Third World.

Law is the rich nations' approach to resolving conflict. But applying systems of legal justice in these circumstances will simply lead to endless conflict while the world slips into catastrophe.

Many money-poor nations have a different system of resolving conflict: look for common ground. The poor want a world economy that is fair. The rich need environmental stability. The rich minority nations must humbly ask the majority nations to share the damage they have caused. The money-poor nations are asking for confrontation to be put behind us and – please – stop thinking of everything in terms of money and military bully-power. We must work together to save the world from environmental catastrophe and to eradicate hunger.

Free Trade

comparative advantage for corporations

Tatu Museyni cultivates coffee in Tanzania. The price she gets for coffee has halved over the 20-year period since 1980. It halved again in the last two years. Her income is now down to $40 for the whole year. Her children no longer have education and the family faces starvation. Even worse is in store. Researchers in Hawaii have developed GM coffee plants whose beans all ripen at the same time making it possible to pick by machine. When machines replace pickers, Tatu Museyni and 25 million others may have no income at all. But on the other side of the planet, Nestlé boasts in its annual report, **"Thanks to favorable commodity prices profits have reached a record high."**

The price of other commodities also dropped between 1980 and 1997: sugar down 73%, cocoa down 58%, rubber down 52%, rice down 51%, cotton down 36%, copper down 30%. And prices continue to drop. The reason is obvious. A multinational corporation (MNC) will move its operation to where it can get coffee cheapest, say Brazil. If India drops its price, it will move there. If Brazil then drops its price further, it will move there. It is a race to the bottom. This is free trade, otherwise known as market fundamentalism and corporate capitalism.

Consider a company that buys land from farmers in Kenya to grow flowers for sale in New York. Previously self-respecting farmers become dependent laborers, and land that once fed the rural community now provides luxuries for the rich. But then the company may suddenly transfer its operations to Uganda if labor is cheaper there. Rich countries get cheap luxuries; the company makes a good profit, GDP in Uganda rises and their business elite benefits. But, also in Uganda, the poor become even poorer and, in the US, flower-growers lose their business.

Three-quarters of all international trade is now in the hands of MNCs. It enables them to drive down the cost of commodities (products in US shops become cheaper), to have more customers (they can extract profit from poor countries) and to locate their facilities where environmental and labor standards are low (if unions demand proper wages the MNC can move elsewhere). They dismiss horrific living and working conditions and early death as "commercial reality."

The concept of corporate free trade is forever associated with the *laissez-faire* monetarist policies of President Reagan and Margaret Thatcher. John Maynard Keynes had a more enlightened and humane approach: "I sympathize with those who would minimize entanglement between nations. Ideas, knowledge, art, hospitality, travel – these are the things that should of their nature be international. But let goods be homespun whenever reasonably possible and, above all, let finance be primarily national."

Corporate free trade has seen a period of increasing poverty and social disintegration, alienation, breakdown of democracy, violent insurgency groups, environmental degradation and new diseases. A third of the world will soon be living in city slums. Already more people are hungry than ever before.

For the global elite, free trade corporate capitalism is a fundamentalist religion, impervious to reason, evidence, or compassion. Like communism before it. The religion has a central dogma that an "invisible hand" will somehow result in the poor benefiting from the excessive wealth of the rich. World leaders should open their eyes to the ghastly results of their fanaticism but all they can offer is, "Hold tight, it will be all right in the end." This was Stalin's story. In the end, the oppressive regime collapses, leaving social and financial chaos.

There is support for free trade among businessmen and governments in poor countries. The World Bank's data showed that free trade raised their GDP. To demonstrate this it commissioned research. But to its distress the report, "The Simultaneous Evolution of Growth and Inequality," showed that free trade benefits only the elite in poor countries but sinks the lower 40% into deeper poverty. These findings did not support the fundamentalist beliefs of the Bank's hierarchy and so were deleted. The editor of the World Development Report, professor Ravi Kanbur, resigned in disgust.

- The US and Europe promote their own produce with massive subsidies, export credits and tariffs and prevent others using their knowledge. Poor countries are forced, with sanctions and withdrawal of aid, to open their markets. This prevents them from adding value within their borders and so are prevented from climbing out of poverty.

I think we had more freedom before

- In the last two years 500 export factories in Mexico have closed, firing 218,000 workers. Their $1.26 an hour wage was too high. Alcoa now produces auto parts for the US in Nicaragua where workers get 40 cents an hour. Now these workers have been told to work for even less or the jobs will go to China where workers will get 16 cents an hour, work 16 hours a day, 7 days a week with no health insurance and no pension. This is market fundamentalism in action.

- **Comparative Advantage**: Two countries trade hats and oranges. One is good at making hats; the other grows oranges. A country benefits from selling what it grows or makes better or more cheaply than other countries. This theory of "comparative advantage" is taught to economics students as the underlying principle of trade and is trotted out by economists and politicians as one of the benefits of the free trade religion. It made sense when Adam Smith formulated the theory in the 18th century, but it has been turned on its head. Now that capital and investment are mobile, the rich buy up profitable land or businesses in poor countries. With global corporate free trade, a poor country *loses* its only advantage.

The law of comparative advantage now reads: **"multinational corporations benefit by taking over those things that a country grows or makes better or more cheaply than other countries."**

Hungry for Trade
John Madeley 2000

The WTO

a better way is possible

After World War II the US had emerged as the only superpower and President Truman could have dictated international policy. Instead he decided it was in the country's long-term interest to cooperate. The Marshall plan helped European reconstruction and the General Agreement on Tariffs and Trade (GATT) sought a level playing field that would protect the weak.

The WTO grew out of the GATT but things have gone disastrously wrong. After 55 years half the world's population lives on less than $2 a day and half those on less than $1. Around the world poor farmers are committing suicide because trade rules are destroying their livelihood.

The WTO has been hijacked by corporate interests and its purpose is now to promote trade. Period. There is no higher authority, and corporate free trade, "market fundamentalism," has become the governing principle of the world. The WTO - deliberately - now ensures that the weak cannot protect themselves against the strong.

The WTO aims to weaken national government. A strong government in a poor country might discourage the export of raw commodities so that value can be added to them within its own borders. It might also exclude imports that had been cheapened unfairly by subsidies in rich countries. These are ways in which the poor can climb out of poverty. But they are forbidden.

Under WTO rules governments are not allowed to favor local firms, prevent foreigners having a controlling interest in local companies, favor trade partners or subsidize domestic industry (though it has not prevented massive US and EU subsidies to their farmers and exporters). Governments are not allowed to interfere in a market to pursue social objectives such as racial, ethnic or gender equality, nor to favor friendly countries that might have special needs. The rules work to the advantage of corporations which benefit from economies of scale, which can undercut to capture a market, which are immune to local consumer feedback and which can shift their production at short notice to countries with lower wages and fewer environmental or labor regulations. In all these fields manufacturers, suppliers, retailers and farmers in poor countries are at a disadvantage. It is hardly believable but the WTO sets maximum, not minimum, standards for environmental protection.

Trade spreads films, music, language, business methods and attitudes. The WTO is therefore establishing a single culture, globally. But modern communications make everyone aware of stark and growing inequality, disintegrating societies, collapsed economies, violent reaction from disaffected groups and environmental crises.

The WTO claims to arrive at decisions by consensus. The Quad (Canada, EU, Japan and the US) sets the agenda and detailed wording of its decisions are hammered out in Green Room meetings to which awkward delegates are excluded or forcibly ejected. Green Room decisions are then presented as "consensus" and binding on all. Objectors are often neutralized by phone calls from heads of state to the delegate's government (George Bush made fifteen such calls during the Cancun meeting). So formal disputes have largely been limited to differences among the rich.

For the notorious WTO meeting in Seattle in 1999 the chair was the head of the US delegation and delegates from poor countries were sidelined and ignored both in preparations and during the conference. At the following Doha meeting, Europe and the US undertook to reduce subsidies and both reneged on their promises. Subsequent meetings were held in remote locations behind police barriers, vividly demonstrating that the WTO had lost legitimacy.

By September 2003 the majority nations had had enough. The EU was responsible for the collapse of the Cancún talks by refusing to withdraw new issues. The US had anyway decided to deal unilaterally – "WTO or no WTO we plan to do just what suits us" said one US ambassador – indicating that it would continue to use the old imperial tool of "divide and rule."

But the majority nations demonstrated at Cancún that by working together they can take control of the future. Either the WTO will be made democratic or a new body will be set up with population-based majority voting.

The primary objective of a world trade organization should be to encourage and enable the poor to add value to their commodities and crops within their own communities and boundaries. Countries with poor communities must be *encouraged* to use appropriate subsidies, tariffs and import quotas, but for these restrictive measures to be reduced on a **sliding scale** as their economies strengthen. Rich countries should remove all fiscal measures that promote exports or benefit established companies, but not those measures that help small businesses to serve a local market.

This policy was pioneered by the American Founders in their struggle against the tyranny of British companies and led to Alexander Hamilton's "infant industry protection" policy that regulated US trade from 1789.

TRIPs (Trade Related Intellectual Property Rights)

The TRIPs agreement protects patent rights – almost all of which belong to western companies – and countries are no longer allowed to make patent regulations to suit domestic conditions. The TRIPs agreement is a restriction on trade in favor of corporations, a blatant infringement of the WTO's own free trade agenda. It limits the use and development of knowledge, paralyzes scientific research, benefits only the rich, prevents technology-transfer to the poor, denies affordable access to life-saving medicines, limits farmers' use of traditional knowledge, reduces bio-diversity and even allows nature's life forms to become private property.

The Age of Consent
George Monbiot 2003

The GATS

exit democracy and peace

Negotiations for the General Agreement on Trade in Services (GATS) should be abandoned until an independent assessment has been made.

The GATS came into being in 1995 following pressure from multinational corporations (MNCs), but details are still being worked out. The European Commission admits: "the GATS is first and foremost an instrument for the benefit of business." There has been widespread protest in poor countries as they see their services taken over by foreign companies but, after nine years, the WTO has carried out no independent research into its effects.

I thought they believed in democracy

The rich say that many poor countries have not been able to afford adequate investment in their service sectors so multinational corporations will help them with both capital and expertise – the rose-tinted view of privatization.

Under the GATS, national governments must submit to certain rules when handing over services for privatization. Countries are free to decide which sectors they will subject to GATS rules though the aim is for all to be privatized in the end. **The GATS enables corporations to invest, acquire land and take over essential services in any country that signs up.**

Health, education, electricity, water, sanitation, post, telecommunications, transport, banks, investment, insurance, radio, television, film, garbage collection, setting up of retail stores, construction, tourism, land – they all fall within the orbit of the GATS. The preamble to the GATS has many statements to reassure its critics. But the preamble is not legally binding. The rest of the document is legally binding and it specifies that governments must not regulate in any ways that are a hindrance to trade. Even a government's right to hold back essential services is being eroded in current discussions. Article V1.4, the "Necessity Test," makes the GATS Dispute Panel arbiter with veto powers over whether a government's regulations conflict with free trade imperatives. Bureaucrats that meet in secret determine what a government can or can't do. National parliaments are thus demoted to mere advisory bodies within their own countries.

Many consider the GATS to be the most evil and dangerous of all interventions by the WTO in the affairs of poor countries. If politicians have lost power over economic measures they are left with only the manipulation of prejudice to catch votes.

At the WTO 2001 meeting in Doha the Quad listened to delegates from poor countries but the GATS magicians kept vital cards up their sleeves:

• Trick 1

The process is irreversible. If a government or a dictator has privatized a service and listed it to the GATS, such as railways, health or water, a subsequent democratically elected government cannot take it out.

• Trick 2

The regulation of trade is enforced by sanctions. Sanctions by poor nations are a scarcely noticed pinprick for the rich. Sanctions and withdrawal of assistance by rich nations can destroy the poor.

• Trick 3

Only rich-nation businesses benefit. One cannot conceive of companies from poor countries bidding for health, education or water services in the US or France or the UK.

• Trick 4

A country is not free to encourage its own expertise and indigenous culture by, for example, limiting the number of foreign citizens working as architects.

• Trick 5

Awkward negotiators from poor countries are withdrawn following a phone call to their governments from a powerful country.

• Trick 6

A poor country may be persuaded to bargain away a particular service because it is desperate to retain, say, favorable tariffs on textiles.

• Trick 7

It is virtually impossible for indigenous

companies to compete with MNCs – either to get started or to survive.

- **Trick 8**

A country may not realize what it is letting itself in for. Having listed the tourism, for example, it may not be able to limit the number of hotels provided by foreign companies at an historic site.

- **Trick 9**

Negotiators from poor countries are frequently excluded from critical meetings.

- **Trick 10**

Countries cannot insist on foreign firms working jointly with domestic firms, or that a proportion of shareholders be domestic, or that some assets be held within the country to cover liability for damages.

The General Agreement on Trade in Services signals the end of national government. When investment is open to all comers, when financial services are run by global banks, when local trade cannot be given preference, when the local language and culture cannot be protected, when governments are forced to sell essential services to multinational corporations, when even the land of a poor country can be appropriated by the rich – what is left? Exit democracy.

GATS for violence and starvation

Democracy brought an end to famines in India in 1947. The new government gave priority to livelihoods, food security, land distribution and it imposed a ceiling on land ownership. In 2002 famine returned to six states in spite of overflowing grain stores. Why? The World Bank had forced India to dismantle its humanitarian structures in favor of trade and commerce. Without democratic influence over economic content, politicians can now only bid for votes on the basis of prejudice, fear and hate. Communal violence, cross-border aggression and starvation are back.

The IMF is un-democratic and its proceedings are as opaque as crude oil.
- World population: **India: 17% US: 4%**
- Share of votes: **India: 2% US: 18%**

Wealth-based voting was rejected by the American Founders and was abolished in Britain in 1832.

The Best Democracy Money Can Buy
Greg Palast 2003

Argentina

Argentina was the poster-child for World Bank and IMF policies during the 1990s. The peso was linked to the dollar. Everything from oil to water was privatized.

At the first sign of economic trouble the global companies transferred their loot abroad. Wealthy individuals who saw what was happening also hastily converted their pesos to dollars. Citizens lost jobs, savings, dignity and many now scour refuse heaps for half-eaten bread. One of the world's richest countries has joined the ranks of the poorest. Argentina shows what lies in store for countries that adopt the IMF's neo-liberal agenda. Its government is begging for loans. It should be demanding reparations.

The US is the world's biggest exporter of cotton and its farmers receive $4bn subsidies for producing $3bn-worth of cotton. Their cheap cotton exports are bankrupting growers in poor countries.

FTAA (Free Trade Area of the Americas)

This proposal is for an extension of NAFTA and is similar to the GATS. It would enable corporations to compete for publicly funded services currently provided by governments, from healthcare and education to social security, culture and environmental protection. There have been protests throughout the Americas at this threat to democracy. In September 2002, ten million people from nearly 4,000 municipalities throughout Brazil responded to the question, "Should the Brazilian government sign the FTAA treaty?" A resounding 98% said "NO!"

"Our objective with the FTAA is to guarantee control **for North American businesses** over a territory which stretches from the Arctic to the Antarctic - free access over the entire hemisphere, without any difficulty or obstacle, for **our** products, services, technology and capital."

Colin Powell, US Secretary of State

Water Denied

crimes against humanity

Control of water-distribution in a water-stressed country is effectively control of a populace. Denying water rights to the poor is a crime against humanity.

Water is more essential than oil. Bottled water is more expensive than gasoline. So, as water becomes scarce, excitement among corporations and their CEOs is palpable. Companies are building massive pipelines, giant sealed water bags and supertankers to transport water to those that can pay high prices. Western governments are bullying poor nations to surrender their water to the GATS. The World Bank wants grandiose water schemes and says that the private sector must be involved, **though any child can work out that "the market" prefers the rich to the poor.**

We all have a right to water, particularly to the rain that falls on our community. But water is being stolen. If the fundamentalist devotees of free trade cannot admit to these simple facts they should look at what is actually happening. The examples below are just a drop in the ocean.

⚑ The World Bank pressured the Bolivian Government to privatize water companies, and a subsidiary of the US giant Bechtel gained a 40-year concession. The poor faced water-rate increases of $20 a month (the minimum wage is $100 a month). Even collecting rainwater in rooftop tanks became illegal without a permit. In la Cochabamba, streets were barricaded, 30,000 protestors seized the central square and police fired on the crowd, which then swelled to 80,000. Bechtel left the country. A coalition of people's activists has taken over the water company (impossible if the GATS rules had already been in force). A few days later Wolfensohn, Director of the World Bank, had learned nothing from the experience, claiming that Bolivia still needed "a proper system of charging," and that there was no option but to pay international prices for a valuable resource. But the fight is not over. Bechtel is suing the Bolivian government and the government is harassing the coalition of activists.

In India a local tribe had always used water from an artificial *tanka* in the forest but it could not prove ownership. The government sold the *tanka* to Pepsi who fenced it and kept the tribe out. Pepsi even tried to prevent the tribe from storing rainwater from their roofs.

South Korea is using virtually all its available water. A detailed World Bank study predicted that withdrawals for industry and housing will increase and water for agriculture will decline from 13 to 7 billion tons by 2025.

Taxpayers are helping the state of Orissa in India to privatize water distribution through an ill-conceived aid project. Water rates have increased tenfold. Millions will be forced to migrate into city slums. But thousands of small farmers are taking control of the water services themselves through non-violent direct action.

A subsidiary of the French company Générale des Eaux was granted a 30-year contract to deliver water in a rural province of Argentina. Water rates doubled and the water turned brown. Customers forced the company out of the province by refusing to pay their bills.

In another incident in Argentina a water company behaved properly. It subsidized poor areas by charging wealthier customers a "solidarity tax." The wealthier customers used the courts to get the surcharge ruled illegal. Under the GATS, it is impossible for a company to pursue socially desirable policies.

In 1995, water privatization in Puerto Rico left poor people without water while US military bases and tourist resorts remained well supplied.

The World Bank withheld debt relief and funds from Ghana until water services were privatized. The Ghana National Coalition Against the Privatization of Water was formed.

A Coca-Cola plant in Kerala uses so much water that it has deprived 2,000 people of their supply. Farmers were pleased to accept the company's waste as fertilizer; but after three years it was found to contain toxic metals including cadmium that causes kidney failure and lead which is particularly dangerous for pregnant women. These poisons have spread to the water supply and the food chain. Coca-Cola in Delhi was found to contain pesticides and it was decided that Coke and Pepsi should no longer be served in parliament.

Does someone own my water?

When companies fail to milk the poor they pull out. Suez abandoned Manila. Saur left Mozambique. Vivendi is moving out of all poor countries. The poor are left to rebuild the ruins.

Almost all parts of India have adequate rainfall if the water can be stored where it falls and fed into aquifers. "Rainwater harvesting" is the most efficient water conservation and distribution system. Tens of thousands of *tanka* reservoirs can be seen throughout the sub-continent testifying to this tradition. Communities consider that they have a right to the rain that lands on their fields. But the WTO has re-defined water a "need" not a "right" and, in its language, a "need" is something that can be bought and sold – to corporations. In cities, participatory water management has proved a viable alternative both to privatization and to traditional bureaucratic public models. "Through social control, democracy and transparency, people push us to be more efficient," says Carlos Todeschini of Porto Alegre, Brazil.

Abhorrent policies result when money takes precedence over civilized values. Lawrence Summers, when Chief Economist to the World Bank, noted that health-impairing and death-inducing pollution costs are lower in poor countries. "I think," he wrote "the economic logic behind dumping a load of toxic waste in the lowest-wage countries is impeccable and we should face up to that." 80% of US electronic waste is sent to Asia. 100,000 Chinese villagers are paid $1.50 a day to salvage valuable parts and dump the rest, unaware of the chemical hazards. Much of it goes into irrigation canals – and gets into the food chain and aquifers.

www.corporateeurope.org/water

The First MNC

little has changed

The East India Company was the first major shareholder-owned multinational company (MNC). It found India rich and left it poor.

When the company was established in 1600, and for 150 years thereafter, there was nothing England could export that the East wanted to buy. Spices, textiles and luxury goods sailed west. Only money sailed east. **It was the ability to acquire land and control government services that made the fortunes of the Company – and broke India.**

As the mighty and opulent Mughal Empire declined, the Company acquired land beyond its vulnerable trading ports, extorted taxes, manipulated terms of trade in its favor, and built up a private army. In 1757 Robert Clive fought and defeated the Nawab of Bengal. Later, Lord Cornwallis defeated Tipu Sultan in

the south. In both cases, and in many lesser incidents, the Company's executive officers extorted huge ransoms and accumulated unimaginable wealth. This wealth was obtained, not so much from its fashionable society customers back in England, as from suppliers in India, from defeated rulers and from taxes imposed on the populace. India financed its own impoverishment.

Under the Mughals, taxes had been collected through a complex pattern of mutual obligation. This was too complex for the Company. At a stroke the *zamindars*, tax farmers under the Mughals, were transformed into landlords, and Bengal's 20 million smallholders were deprived of all hereditary rights. Just five years after the Company secured control of Bengal in 1765, revenues from the land tax had tripled, beggaring the people. The devastating effects last to this day. These conditions turned one of Bengal's periodic droughts, in 1769, into a full-blown famine. An estimated 10 million people died. But, rather than organize relief efforts the Company actually increased tax collection during the famine. Granaries were locked, and grain was seized by force from the peasants and sold at inflated prices in the cities.

The Company became feared for the brutal enforcements of its monopoly interests. For example, it would cut off the thumbs of weavers found selling cloth to other traders to prevent them ever working again. In rural areas two-thirds of a peasant's income was taken in tax, nearly double that under the Mughals. The Company's performance, through pursuing profit for its shareholders and its chiefs, contrasted starkly with its claim, in the mid-19th century, that it ruled for the moral and material betterment of India. The British government followed the Company's policy. To maintain a monopoly in salt, the rulers made it illegal for Indians to make their own, and consumption of salt was forced even below the minimum prescribed in English jails. This disgraceful control of an essential commodity was only withdrawn after Gandhi's famous Salt March in 1930.

In Britain, so powerful was the Company that attempts to control its affairs could bring down a government. An attempt, led by Edmund Burke, to place the Company's Indian possessions under Parliamentary rule led to the dismissal of the government. The general election that followed was so generously funded by the Company that it secured a compliant parliament - a tenth of the seats were held by "nabobs."

Booty from India created this new class of nabobs, the chief executive officers (CEOs) of Georgian England. The nabobs themselves had no conscience about their wealth. Robert Clive, having extorted a fortune after the battle of Plassey, defended himself at a House of Commons enquiry into suspected corruption, saying that he was "astounded" at his own moderation at not taking more. Only a few dissenting voices, like the Quaker, William Tuke, pointed to the humanitarian disaster that the Company had wrought in India. But the case for reform was overwhelming and in 1784 the India Act transferred executive management to

a Board of Control, answerable to Parliament – a kind of public-private partnership.

Although the Company pretended to a mission to make Indians "useful and happy subjects" the underlying ethics of the public-private partnership remained the same. By the 1850s, just £15,000 was being spent on non-English schools – the military budget was £5 million. Railways were built to speed access of British goods to Indian markets. Mill-made cloth brought from Britain shattered the local village economies, which were based on the integration of agriculture and spinning. The great textile cities of Bengal collapsed. The Governor-General reported, "The misery hardly finds parallel in the history of commerce. The bones of the cotton weavers are bleaching the plains of India."

Indians were worn down by the British dominance, by the unfair trading rules, the crippling taxes, the draining of India's wealth and the contempt in which the majority was held. Retaliation was inevitable. The final insult was forcing *sepoys* to use a rifle cartridge greased with cow and/or pig fat – an outrage to both Hindus and Muslims. Catastrophe struck in 1857. The British could not understand the hatred and called it "mutiny." The massacre of Europeans generated a ferocious bloodlust in English society and led to brutal reprisals. Long-standing plans for increased dominance in all spectrums of Indian life and economy had at last received their justification. In 1858 the East India Company was abolished and direct rule by Queen and Parliament was introduced.

Corporate exploitation followed by an outrage led to Empire. Is history repeating itself?

Loot: Resurgence 210
Nick Robins 2002

Ending Tyranny

the struggle of the American Founders

"I believe in this beautiful country. I have
studied its roots and gloried in the wisdom
of its magnificent Constitution. I have
marveled at the wisdom of its founders and
framers. Generation after generation of
Americans has understood the lofty ideals
that underlie our great republic.
I have been inspired by the story of their
sacrifice and their strength.
But today, I weep for my country."
Senator Robert Byrd
March 2003

Boston's citizens threw $15,000 worth of tea
into the harbor. That was a lot of money in 1773
– and a lot of tea. The tea belonged to the East
India Company, the world's first multinational
corporation (MNC). The British government then
decreed that the harbor would be closed until
the people of Boston reimbursed the Company.
That led to war and rejection of the tyrannical
regulations imposed by a remote government

on behalf of a corporation. Poor countries today
have similar feelings about corporate tyranny.

The American Founders were determined to
prevent any tyranny enslaving the New World,
whether the tyranny of an aristocracy or of
standing armies. But it was trade dominance
that aroused their greatest passions. At the
time, virtually all members of the British
parliament were stockholders with the East
India Company, a tenth had made their
fortunes through the Company, and the
Company funded parliamentary elections
generously. British traders and politicians
prospered through the activities of this first
MNC that was sucking wealth from around the
world. The parallel with today's corporate and
political life in the United States is remarkable.

Measures were introduced in the New World to
prevent corporate tyranny. Corporations
received their charters from individual states

and these were for a limited period, say 20 or 30 years, not in perpetuity. They were only allowed to deal in one commodity, they could not hold stock in other corporations, their property holdings were limited to what was necessary for their business, their headquarters had to be located in the state of their principle business, monopolies had their charges regulated by the state, and all corporate documents were open to the legislature. Any direct or indirect political contribution was treated as a criminal offense. Corporations had their charters removed if the state considered that their activities harmed its people.

Railroad companies had traditionally been referred to as "artificial persons" and when the Fourteenth Amendment gave all "persons" equality before the law they desperately tried to claim that equal rights applied not just to slaves but to them as well. For eighteen years the Supreme Court consistently ruled that corporations did not have the rights of human persons. Then the *Santa Clara County vs. Southern Pacific Railroad* legal case reversed this. Corporate tyranny in the US and throughout the world can be traced to this one case.

Textbooks only quote the headnote of this case, not the detail. It was Thom Hartmann who eventually unearthed the original records in Vermont only to find that the judge had specifically stated that the case did not relate to corporate personhood. The headnote had been written a year after the hearing by the Recorder, a person whose life had been with the railroads. By then the judge was too ill to check the headnote. **Corporate American law is therefore based on a fraud.**

I thought Americans admired their Founders

The second step to corporate tyranny was adoption by the US of the WTO and NAFTA agreements. It is commonly thought that these benefit the United States. Wrong. The agreements deprive US citizens and authorities of the ability to make decisions about their local environment and trade, giving power to foreign bureaucrats who meet in secret. NAFTA's effect on Mexico, where eight million fell from the middle class into poverty, has been well documented. But in the US 2.6 million jobs vanished (many to be replaced by part-time or low-pay service sector jobs), workers lost $28 billion in bonuses, wages fell 28%, working hours increased, environmental laws were overruled and the trade deficit soared. It was only corporations that benefited, through being able to overturn local laws and force down the cost of labor.

All the protections against corporate power sought by the American Founders have been lost. Corporations, with the right to "personhood," now have the right to meet and influence "their" elected representatives. Chemical corporations, quoting the Santa Clara case, claim the right to privacy to prevent uninvited inspections of their toxic sites by the EPA. Corporations have the right to sue an objector at a public meeting for slander, and frequently exercise it, not hoping to win but to intimidate with the threat of legal fees. Corporations, asserting the right to free speech, used a perverse advertising campaign to kill off Clinton's proposal for health-care protection to the 40 million uninsured.

Can we return to the values of the American Founders? Reversal of the Santa Clara case would be the first step to subjecting corporations once again to the control of the people. The federal government, each state, each township, could then regulate corporations to the benefit of its citizens and help local economies to flourish once again. Indeed, in California local governments have already passed laws that deny corporations the status of persons, while in Pennsylvania some townships have forbidden corporations from owning or controlling farms in their communities.

Huge questions remain. Deprived of personhood the corporations would still have bully-power.

But, because of its unique constitution, the United States could lead the world once again towards government of the people by the people, for the people (not by and for corporations).

Contributions for 1996 elections:

96% of Americans	$0
the next 3.75%	up to $200
Top 500 corporations	over $500,000 *each*

In 1998 elections the top 500 corporations gave $660,000,000 to candidates.

Unequal Protection
Thom Hartmann 2002

The Empire

full spectrum absurdity

Were four thousand years of culture more nourishing than fifty years of development?

"All this talk about first we do Afghanistan, then we do Iraq. . . this is entirely the wrong way to go about it. If we just let our vision of the world go forth, and we embrace it entirely and we don't try to piece together clever diplomacy, but just wage total war . . . our children will sing great songs about us years from now."
Richard Perle 2003

The US is the world's only hyperpower. It, with the elite of other countries, is undertaking a vast utopian project to reorganize the world through an economy based on corporate market fundamentalism with common cultural values and western style democracy. Whether they like it or not, all nations must obey the rules.

The Project for the New American Century (PNAC) set out the neo-conservative blueprint for the development of this empire in June 1997. Later, in September 2000, the PNAC said, "The process of transformation, even if it brings revolutionary change, is likely to be a long one, absent some catastrophic and catalyzing event – like a new Pearl Harbor." 9/11 was the defining event that moved US trade dominance into overt empire - in the way that the Indian Mutiny led to the British Empire or the burning of the Reichstag confirmed Hitler's power. **The War on Terror is now a useful definition that enables the US to target any state or group that threatens its interests anywhere in the world.**

The dominant position of the US could have provided peace and stability through cooperation with other nations in a world of weak governments, corrupt global corporations and anti-elite radical groups. Its economy has stimulated trade when most others were stagnant or collapsing. Its communications science has made the world into a global village - though with doubtful cultural value.

But the US has rejected the United Nations' multilateral approach to world affairs. It now acts unilaterally. Examples are legion and include: rejection of the Comprehensive Test Ban Treaty, withdrawal from the ABM Treaty, violating and killing the Biological and Toxic Weapons Convention, rejecting the Small Arms Treaty, rejecting the Land Mines Treaty, rejecting the Clean Energy Plan and refusal to allow UN inspection of its weapons of mass destruction. It has opposed negotiations for preventing the weaponization of space, refused to sign a treaty on the rights of the child, "un-signed" the treaty for an International Criminal Court and ignored the Geneva Convention. The list of rejections goes on: UN Global Program against Business Corruption – Law of the Sea – Economic Espionage – Harmful Tax Competition – Racism – Torture – Discrimination Against Women . . .

The life-blood of the Empire is oil, so the US must have ultimate control over oil production if it is to maintain its global hegemony. The world's biggest oil reserves are in the Middle East, where

the US now has 27 military bases. The Caspian/Gulf zone, sometimes referred to as the 51st state, is at the heart of the land mass that includes China, the Indian sub-continent, Indonesia, the Middle East, Africa, Europe and Russia. Oil and gas pipelines radiate from here. With control of this zone the US can threaten the economy of any country. China is the endgame, being the workshop of the world with a trade surplus that threatens the US economy and with uncertain commitment to the empire project. China is dependent on imports for two-thirds of its energy needs so has a legitimate fear of US rhetoric.

The US now seeks universal hegemony through military power. **The mission statement of its armed forces is "full spectrum dominance,"** dominance on land, dominance of the oceans, dominance of the airwaves and intelligence, and dominance in space. Countries are defined simply as either friends or enemies of the US.

Since the Cold War, the US has reduced its army but increased expenditure on intercontinental-range bombers, the Navy's long-range strike capabilities, forward deployment of supplies especially in the Gulf and the Indian Ocean and counter-insurgency groups in 50 countries.

The US now aims to revolutionize warfare with directed-energy, space-based laser weapons on invulnerable satellites circling the earth that are able to target any stationary or moving object on the ground, in the air, or in space at the speed of light and with extreme precision. "The globalization of the world economy," says the US Space Command's "Vision for 2020" report, "will continue with a widening between haves and have-nots. From space the US would keep those have-nots in line." And the Commander-in-Chief of Space Command said, "We're going to fight *in* space. We're going to fight *from* space and we're going to fight *into* space" (his emphasis).

Friendly states, like Britain, provide moral support and help with bases and intelligence facilities but their interests will be ignored when they are no longer considered useful, as France found when it questioned the legality of the Iraq invasion. Less friendly states suffer economic or military aggression. Poor countries, like the 35 that refused to guarantee the immunity of US citizens from prosecution, have aid terminated.

However, the reliance on overwhelming power and technical wizardry meets untidy human complications. Arming the Northern Alliance against the Taliban in Afghanistan spread arms to warlords that fight the US-imposed government. The opium trade, banned by the Taliban, has escalated. The anticipated welcome to liberators in Iraq turned sour. The empire's ability to conquer has been demonstrated but it has no effective strategy for achieving peace. Fundamentalist groups within the US that equate current developments with Biblical prophesies of Armageddon twist logic to such an extent that peacemaking is seen as contrary to God's will.

But the US position is simple: the strategic interests of the United States are not negotiable and, in the words of George W Bush, **"You are either with us or against us."**

"Fascism should more properly be called corporatism, since it is the merger of state and corporate power."
Benito Mussolini, fascist dictator 1925-43

Imperial Tribute

why the rich are rich and the majority poor

The United States prints dollars, bits of paper or computer entries that don't cost anything to make, for other people to use. This is its biggest export. Professor Thomas Barnett of the US Naval War College admirably explained the process:

"We trade little pieces of paper (our currency, in the form of a trade deficit) for Asia's amazing array of products and services. We are smart enough to know this is patently an unfair deal unless we offer something of great value along with those little pieces of paper. The product is a strong US Pacific fleet, which squares the transaction nicely."

In other words, in exchange for vast quantities of goods and services, the poor get a US military presence. Big deal!

THE TRIBUTE
A third of all goods and services imported by the United States in exchange for computer entries.

What is real wealth? People grow food and flowers, mine minerals, refine them, design and make things, create art and music, run businesses and do all those things that make a country tick. These are the things we want, the things that make for a good life. We provide these things for ourselves in our own countries and trade them with others. They are the things that flow into, and out of, every country. But the US does not export as many of these things as it imports. A third of its exports are simply – dollars.

People want these dollars because they are an international currency. A sheikh gets dollars for the oil he sells the US. He might buy a skyscraper in Sydney as an investment. The Australian, in turn, may buy goods from Japan using the dollars. The Japanese may set up a factory in the US. So the dollars circulate. The more trade there is, the better for the US - people will want more dollars. Countries, also, need a cushion. So there are dollars in central banks, in businesses and in private accounts all around the world.

The amount of these dollars, that have left the US in return for imported goods and services (the net foreign liability), is now huge, $2,500 billion. Yet it would cost only $40 billion a year for 10 years, $400 billion in all, to ensure that everyone in the world had adequate diet, safe water, basic healthcare, adequate sanitation and natal attention.

The ability of the US to export an international currency in exchange for real wealth is one of the main reasons why most of the world is poor and the United States is rich. The poor subsidize the rich.

But the US is on a treadmill. In order to support its standard of living it needs to keep exporting more and more of this international currency. Morgan Stanley, the investment bank, says that it will soon need to be exporting $2 billion a day (the trade deficit). The US is therefore desperate for trade throughout the world to increase so that people will need their currency. It is annoyed with Japan for failing to get its economy going. It is annoyed with Europe for not growing fast enough. It wants poor nations to cultivate cash crops for export, rather than for local use. And it insists that heavily indebted poor countries will only have their debts cancelled if they open their markets to global free trade.

This is not sustainable, so the value of the dollar will fall, is falling, and the US becomes vulnerable. If Saudi Arabia withdrew its investments or if it withheld the flow of oil, the US economy would crash. Maybe this is why, following 9/11, the US attacked a defenseless country rather than the one from which hijackers and finance came.

This perverse financial arrangement started at Bretton Woods in 1944 when the US was given the right to have its dollars considered the equivalent of gold in the reserves of central banks throughout the world. General de Gaulle of France called it "an exorbitant privilege" though the value of the privilege was limited while the dollar was linked to gold. But in 1971 President Nixon broke this link and the privilege is now unconstrained. Europe would like its own strong currency so that it can pick up the loot as the dollar drops. It is difficult to understand how such an iniquitous arrangement has been allowed to run for half a century and is still in place.

John Maynard Keynes said at the Bretton Woods conference that **there should be a currency for trade between nations that is independent of any national currency.** He was overruled. Fifty-eight years on, his suggestion should be revisited. An independent international currency need not be dependent on the consent of the US. The process could start with a few countries, say the former UK Commonwealth or the G21, exchanging products using their own invented international currency. As dollar reserves are released, the advantages would become obvious and others would join in.

If we are witnessing the last desperate contortions of an insolvent empire, what next? The future may lie with the Japan/China axis if it can learn to live within a finite planet. Western politicians regard these two nations as enemies of each other. Flat-earth economists look only at the financial sector and consider Japan a basket case. They fail to see its real successes: its high real investment in key modern industry and infrastructure, its trading surpluses, its high savings rate and its rising living standards. India, with its expertise in a growing service sector, could reinforce the region. Malaysia contributes economic sophistication.

China and Japan's massive foreign assets have kept the US economy afloat so far but, if the eastern economies with half the world's population continue to grow, these assets would be transferred. Japan already assumes that China will be the world's biggest economy and is lowering both import and export trade barriers both ways while keeping them high to others.

Aid

Keynes would have regarded the necessity for aid, had it started in his lifetime, as a sign that the global financial architecture had failed. The introduction of an international currency, not tied to any national currency, would be the most effective first step towards reducing poverty in the world.

Biomimicry

science's exciting new frontier

A certain spider *(nephila clavipes)* makes six different kinds of silk. When hit by a fast-flying moth it can stretch by 40% and then return to its original size unaffected. The US army is researching spiders because their web is stronger and has better energy-absorbing elasticity than Kevlar, a fiber so tough that it can stop bullets. Kevlar is made by pouring petroleum-derived molecules into a pressurized vat of concentrated sulfuric acid to be boiled at many hundred degrees Fahrenheit and then subjected to high pressures. A spider just needs a fly or two.

A hummingbird can fly forwards, backwards, up, down or sideways, with equal ease. It can fly across the Gulf of Mexico on a fraction of an ounce of fuel and, when picking up its fuel, fertilize the flower to provide more fuel in future. Compare the resources consumed and the damage done in building and flying a Chinook. We have a lot to learn. Architects are fascinated by the ability of termites to build structures that maintain a constant temperature in their interior chambers. Computer scientists think that ants might help them with parallel, rather than linear, computers. And if we want a lot of hydrogen, now that we are entering the hydrogen economy, nature makes it all the time with the help of an enzyme. Can we mimic nature?

Material scientists boast about the incredibly high pressure they can achieve to make hard ceramics (an oyster beats them), high temperatures for special steels, and elaborate chemical processes (they can't match a mollusk for underwater glue). Theirs is outmoded Stone Age technology – *heat, beat and treat*. Some agricultural researchers are noticing that nature does not use the plough. Plants naturally grow as mixed communities, with seldom fewer than four species together.

Nature favors perennials. Even organic farming still has a lot to learn from nature.

Then energy. Almost all our energy comes from the sun – via plants. Coal and oil are embodied energy captured millions of years ago. **Duckweed captures solar energy with 95% efficiency, many times better than man-made photovoltaic cells.** A leaf takes in water and carbon dioxide, the carbon forms sugars that store the sun's energy and oxygen is given off. But how? Scientists say "membrane potential" -where a positive electron is moved to one side of a membrane and a negative electron to the other. If we could mimic that, and if the process needed no more inputs than stagnant water and daylight and produced no more waste than duckweed produces, we would have a truly sustainable energy source. The scientific race is on.

Genetic knowledge is the biggest recent leap forward, but experience suggests that immensely powerful new technologies should be approached with caution. Fossil fuels, while making possible the lifestyle we value, have been largely responsible for overpopulation, degraded land, changing climate, rising sea levels and resource wars. We are on a unique planet on which life has evolved in a unique way and we must live within the planet's unique patterns. In three billion years, all sorts of diverse properties have been combined, like crystals and membranes in a spider's web. But not everything. Boundaries have been erected, and it is these boundaries that we are excitedly destroying when taking genetic inheritance into our own hands. The most pressing need is to recognize how use can be made of genetic knowledge without upsetting three billion years of evolution. Novel genes may cause havoc as they move vertically and horizontally through future generations of bacteria, plants, animals, and ourselves.

Biomimicry introduces an era based not on what we can extract from nature, but on what we can learn from her. "Doing it nature's way" has the potential to change the way we educate our children, grow food, make materials, harness energy, heal ourselves, store information and conduct business.

Biomimicry
Janine M Benyus 1997

Soil

storehouse of nutrition

In a spoonful of healthy soil there can be a billion organisms from over 10,000 species. Soil is alive – a mineral, animal and vegetable kingdom under the more visible green canopy of nature. It is the storehouse from which plants extract resources and in which they lay down reserves for the future. Scientific understanding of the way plants use the soil seems to be in its infancy, comparable with physics about a hundred years ago when the realization dawned that there was power in the atomic structure of matter.

A single worm can shift 30 tons of earth in its lifetime. Together with insects the worms aerate the ground, drag rotting material down, digest and excrete it, and begin the process of transforming it into humus. Bacteria, fungi, protozoa and algae then get to work and release the nutrients from the humus as minerals, proteins, carbohydrates and sugars.

Plants extract up to 10,000 naturally occurring compounds known as phytonutrients, which nourish them and protect them from pests and disease.

Then there are mycorrhizae, the white fuzz around the plant roots. Mycorrhizae create a living bridge between the plant, the minerals and the microbial community with organic acids in a two-way process. The plants grow and feed insects and other animals. Then falling plant matter and manure is deposited onto the soil, returning the nutrients that the plants extracted. The cycle of life continues and the topsoil becomes richer. Beware of simplistic statements about our food. Sir Albert Howard, one of the first agricultural scientists to appreciate these systems, came to the conclusion that **"the maintenance of the fertility of the soil is the first condition of any system of permanent agriculture."**

Through most of history, farmers have worked with nature, noticing when things go wrong. Sometimes things went badly wrong. In the Fertile Crescent of the Middle East, rising salts destroyed both the crops and the empires that depended on them. In the Roman Empire, returning generals invested wealth from conquests in farms and ran them with managers and slaves. Farmers became soldiers and their indigenous knowledge went with them. Rome declined and fell and the fertility of peninsular Italy has never recovered. Things went badly wrong last century with industrial farming when wealth again concentrated land and control into a few hands. Traditional and local knowledge was lost; machines assumed the role that slaves had in Rome and the world lost much of its topsoil.

The quality of what remains is deteriorating. Since 1940, the iron in spinach has dropped 60%, broccoli has lost 75% of its calcium, carrots have lost 75% of their magnesium, watercress has lost 93% of its copper – the list goes on. Selenium, potassium, phosphorus, iodine, molybdenum, sodium, chromium, manganese – over 20 trace minerals have been analyzed and the quantity of all of them has been diminishing in a wide variety of fruit and vegetables. It is these minerals that help keep plants healthy, that are essential constituents of our bones, teeth, muscle, soft tissue, blood and nerve cells - minerals that are involved in almost all the body's metabolic processes.

With increasing knowledge one would expect agricultural science to have found ways to correct this trend. But the reverse is happening.

The Little Food Book
Craig Sams 2003

Feeding the World

there is no shortage of money or food

The National Academy of Sciences and the Royal Society, the leading scientific bodies of the US and Britain, said in their joint report of July 2000:

"Modern agriculture is intrinsically destructive of the environment. It is particularly destructive of biological diversity. The widespread application of conventional agricultural technologies such as herbicides, pesticides, fertilizers and tillage has resulted in severe environmental damage in many parts of the world."

At the beginning of the last century, most food was grown and distributed locally. At its end, just 20 multinational corporations dominated a food trade that had gone international. In spite of this, well over half the population in poor countries is still in farming families. So the most effective way to feed the majority of the world's poor is to ensure that small farmers have good access to thriving local markets, that they are not forced off the land, do not have to buy seed, are helped with appropriate research and technology and have control over water in their locality. The price they get for their produce must not be undercut by subsidized imports.

Two-fifths of the world population is malnourished. Half of these are hungry. The other half eats too much unhealthy junk food. This is hardly surprising because most nutrition research is now funded by and for industrial agribusiness. It is in their interest to develop and advertise food that will appeal to the rich not the poor, to encourage growing for export rather than to meet local needs, to make farmers dependent on purchased seed, chemicals, and machinery. It is not in their interest to make farmers self-sufficient.

There is no world shortage of food. The US has huge surpluses. India exports grain and meat to wealthy countries and even considers dumping excess grain at sea. Poverty, not lack of food, in an increasingly wealthy and unequal world, condemns a billion people to live daily with the horror that they and their families may not have enough, or anything, to eat.

Aid from the rich to the poor is inadequate. Whereas Denmark donates over 1% of its GDP in aid the US gives only 0.1%. But aid is also an admission of failure; it would not be necessary, other than for disaster-relief, if the world had a fair economic system.

The twenty companies that dominate the food trade would like us to believe that increasingly specialized crops, genetic engineering and new chemicals are required to feed the world.

Risk from chemical farming.

Arable land is constantly being lost because chemicals destroy the quality of soil. Dependence on chemicals puts poor farmers in debt and drives them from the land. Nitrates seep into ground and poison drinking water. Persistent synthetic chemicals (POPs and EDCs) endanger human and animal health and reproduction. The short-term advantage of herbicides and pesticides soon turns into a long-term hazard as resistance develops and the balance of predators is upset.

Risk from monoculture.

Diversity is fundamental to all life. The current use of only high-yield and GM crops is a tendency towards monoculture, so is contrary to sound science. During the 20th century three-quarters of the genetic diversity of agricultural crops was lost, 100,000 varieties of rice have been reduced to a few dozen and three-quarters of the world's rice now descends from a single plant. The Irish experience with potatoes was just one warning that monoculture is the herald of starvation.

Risk from specialization.

Crops are increasingly produced where they can be grown cheaply and transported around the world. This creates single-crop farming which makes the whole crop vulnerable. When farmers grow a cash crop for export rather than local sustenance, the local economy is made hostage to the global economy.

Risk from dependence.

Industrial agriculture is dependent on phosphorus, a limited resource, for fertilizer and on oil for chemicals, distribution and machinery. Oil is also a limited resource. It will escalate in price and is causing climate uncertainty which may make it impossible for a farmer to know when to plant, which plants will survive to harvest or to know when that harvest will be.

Risks from reliance on imports.

When societies take unnecessary supplies from abroad, the security of their food is put at risk. Cheap wheat, subsidized by the US, is bankrupting thousands of poor farmers. Britain imports apples and as a result has lost most of its orchards. European dairy products are destroying local production in milk-rich Mongolia. Dutch butter costs less than Kenyan butter in the shops of Nairobi. Imports of food should be restricted to what cannot be grown, or cannot be grown in sufficient quantities, locally.

Risks from productivity.

There are two ways of measuring productivity: first what one person can produce; secondly what one acre can produce. Big business uses the first measure and in poor countries this drives small farmers off the land - to city slums. By 2030 it is predicted that a third of the world will be slumdwellers. What matters is the second measure, productivity per acre. The most productive land is the vegetable plot of an enthusiast who feeds her family and gives or sells to neighbors. She can use the ground for a great variety of produce and maintain its fertility often with just kitchen waste. When she grows for a local market some is wasted because she has to have a certain minimum of each crop and some is not sold. Next, growing for shops is less productive because the shop requires uniform larger batches and it throws more away. Finally large-scale production for supermarkets is the least productive per acre since anything not absolutely uniform must be discarded.

The present surplus could be used to abandon industrial agriculture and establish sustainable, local, mixed, science-based farming methods that are organic – with good access to shops.

Cereal produced by industrial agriculture requires seven times as much energy input as traditional mixed farming.

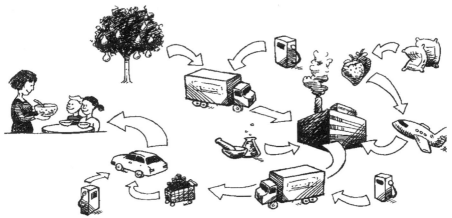

The tendency to eat more meat is not sustainable. Ten acres can produce either one ton of beef or between 50 and 100 tons of grains, pulses and vegetables.

India exports 75% of the meat it produces and 37% of its arable land has been diverted for the export of animal feed; yet many Indians are hungry.

With intensified use of pesticides in the US the proportion of crops lost to pests has increased from 32% to 37%. Many pests are now resistant to pesticides.

The Great Fruit Salad Story
Fruit salad goes from tree to table by modern, industrial, unsafe, irrational, quasi-scientific, bureaucratically controlled, fuel-inefficient, economically-perverse farming and distribution methods. An indigenous person expends one calorie to get four calories of food. How much energy is used to get four calories of food now?

The Doubly Green Revolution
Gordon Conway 1999

Organics

respecting nature's way

In 1896 Professor Shaler of Harvard University said: "If mankind can't devise and enforce ways of dealing with the earth which will preserve the source of life, we must look forward to a time when our kind, having wasted its inheritance, will fade from the earth." **Since then America has lost half its topsoil.** It can take at least 500 years to develop an inch of topsoil, so the damage is immense.

Eve Balfour, one of the founders of the Soil Association, a woman in the man's world of mid-century agriculture, saw land as a fountain of energy flowing through a circuit of soils, plants and animals. She defined food as living channels that conduct energy upwards, while death and decay return it to the soil. In 1946 she warned that chemicals like DDT would get into, and work their way up, the food chain. This hazard took 50 years to enter the consciousness of administrators.

Twenty years later a scientist, Rachel Carson, wrote Silent Spring and thus helped to start the ecology movement. It warned of a countryside depopulated, devoid of wildlife and devastated by chemicals. Since then, man's attack on nature has intensified to all-out warfare as tractors compact the soil, herbicides destroy the humus, and pesticides kill micro-organisms, insects and natural predators. If agricultural scientists claim credit for improving the yield of specialized crops then they must also admit responsibility for the tragic results of their myopic interventions. At last, at the opening of a new century, the general public is beginning to realize that our inheritance hangs in the balance.

When a farmer changes from chemical to organic farming, the productivity of his farm immediately drops. This is because chemicals have degraded the soil and killed predators. It takes several years to restore the land to full productivity. Comparative studies for corn, wheat, soybeans and tomatoes have found that fully restored organic fields average between 94 and 100% of the yields of nearby conventional crops. But these comparisons ignore size. The smaller the farm, the greater the yield. A small organic farm with mixed crops, intercropping and animals can have a considerably higher yield than large conventional farms. This is the farming that can do most to feed the world and, not only in poor countries, can prevent farmers from being uprooted.

Eve Balfour linked the land with our mental, social and spiritual experience. She insisted that contact between town and country should be encouraged, and that farming should be an integral aspect of education. It will be many years before this vision of truly organic sustenance can become a reality. It bears very little resemblance to "supermarket organic" which uses airplanes and lorries to move produce around world.

Over the past decade sales of organic produce have increased at an annual rate of 20%, the fastest growing sector in US agriculture. It is 95% as productive as conventional farming in spite of the billions of dollars the USDA has lavished on chemical farming. If a fraction of that support were given to organic research, it would be more productive.

But why is organic produce more expensive? Because the market is distorted. Market fundamentalism does not apply when agribusiness can influence federal bodies.

Society, not the farmers, carries the costs of pesticide contamination, polluted water and other health, chemical-management and environmental impacts associated with industrial farming - not to mention the $20 billion annual subsidies of the 2002 Farm Security Act. Organic food shows what the true cost of food would be in an undistorted economy.

Cuba, through necessity, provides some interesting lessons. The trade embargo and the collapse of the former Soviet Union meant that it could not import fertilizers, pesticides or fuel for tractors. Most of the food in Cuba is grown in *huertos*, private urban plots of less than a quarter of an acre, and sold from stalls on pavements, at street corners and under the covered walkways of Havana's elegant, crumbling colonial buildings. Inside the city, chemical fertilizers and pesticides are forbidden and compost is made from household waste.

The government took up the challenge of a local, science-based, low input, sustainable agriculture. Unused city land was given to anyone who wanted to cultivate. Forty percent of state farms became co-operatives while the rest were divided into small units. Thousands of jobs were created, 200,000 in 2001 alone. "The secret is in the high productivity," said Nelso Compagnioni of the Institute for Tropical Agriculture. "Every dollar of produce on a small plot costs 25 cents to produce; as soon as you increase the area you get higher costs. And we have no need for transport; customers collect their food on the way home from work."

What will happen when sanctions are lifted? Will neo-liberalism destroy it all? "There will be tough negotiations," says Mavis Alvares, director of the Association of Small Farmers, "it simply isn't the policy of the government to have cheap imported food. We've put an immense educational effort into sustainability."

Sloppy farming practice in Britain resulted in mad cow disease and the horrors of CJD, the human equivalent. However, organic farmers had not been allowed to feed animal protein to ruminants since 1983, three years before the first case. There have been no recorded cases of BSE in any animal born and reared organically.

Organic farming bans estrogen-mimicking pesticides (POPs and EDCs). Irradiation and genetically modified organisms are not allowed. In contrast, non-organic food manufacturers can use more than 500 food additives.

The Organic Farm Research Foundation
www.ofrf.org

The Agrochemical Revolution

was it a success?

As the tractors rolled across the Tanzanian savannah Gidam, who is, or was, a nomad, spoke about the *bung'eda* mound inhabited by his father's spirit. "When the mound is ploughed," he said, "the dead man's spirit is lost. You don't know where your father has gone. My children won't know where their grandfather is. We can no longer belong to this land." To agribusiness this is superstition, but to the nomads it was more significant than their forcible eviction, beatings, fines, imprisonment, murder, loss of cattle - the destruction of their way of life that forced them into city slums.

In 1970 the Canadian International Development Agency had a plan for producing food on an enormous scale. New seeds (Canadian), massive machines (Canadian) and chemicals (Canadian) would transform unproductive land. The Tanzanian savannah was turned into prairie farms. This project was capital intensive (Tanzania had no money) and used little labor (which they had). Initially the land was productive, but much of the thin topsoil ran off and blocked waterholes, the fragile ecology was destroyed and flash floods arrived. Tanzania was left with a massive debt. But this story has a hopeful, if fragile, ending. As farms are being abandoned due to soil degradation, the nomads are returning and coaxing the land back to support their herds.

The Agrochemical Revolution in India entailed a massive program of big-dam construction, where canals cut across natural watercourses. Excessive irrigation has lowered water tables, brought salts to the surface, and turned much fertile land into desert. The recent drought in northern India has strengthened a movement to bring *tanka* back into use. These are large square constructions, many of them ancient, found all over the sub-continent. By channeling rainwater into them during the monsoon they provide water through the dry season and replenish the aquifers. It is called "rainwater harvesting."

Nehru, in 1947, had said, "Dams are the temples of modern India," but by 1958 he was having second thoughts: "The idea of big," he said, "is not a good outlook at all. **It is the small irrigation projects, the small industries and the small plants for electric power which will change the face of the country.**" Dams have displaced twenty-five million people in India. There is no spare land, so they drift into city slums. The 155-dam project on the Narmada river and its tributaries will displace another half million people and destroy fertile land. The World Bank kick-started the Narmada project long before any technical and social research had been carried out; when these were eventually done, the Bank pulled out.

In Indonesia in the late 1970s, vast areas were planted with a single variety of rice. These crops, which were sprayed with pesticide, were devastated by a pest called the brown plant-hopper. In fields just a few yards away, where pesticides had not been used, the natural predators of the plant-hopper flourished and healthy rice continued to grow.

Over thousands of years, humans have gradually improved the variety, yield and nutritional value of crops. Then, since 1950, the Agrochemical Revolution (inappropriately known as the Green Revolution) made a dramatic increase in production. More food allowed the world population to increase exponentially. But it has left a legacy of degraded soil, drained aquifers, silting dams, reduced crop-diversity, depopulated countryside, dependence on specialized seed, food transported over great distances, loss of antibiotics, and chemical-resistant pests, bacteria and fungus. A few companies now want to take this further and change the genetic basis of the world's food supply for commercial gain.

Our food supply, more than any other human activity, is dependent on intimate interaction with nature, but the Agrochemical Revolution and industrial farming have destroyed this close contact. Industrial farming should now be abandoned. Our increased scientific knowledge could lead us out of the mess. Research needs to be de-privatized, widened, and applied as never before, to achieve a sustainable food supply while making farming satisfying, secure and available to the maximum number of people.

"Low-tech sustainable agriculture, shunning chemicals in favor of natural pest control, is pushing up crop yields on poor farms across the world by 70% or more.
A new science-based revolution is gaining strength built on real research into what works best on small farms where a billion or more of the world's hungry live and work."
New Scientist 2001

Peasant-led experimentation spread by modern communications can be as effective as high-tech research programs. Some communities have perfected mosquito traps using a tin with holes, a light and a water tray. Elsewhere predators bred in coconuts eat mosquito larvae.

75% of the genetic diversity of agricultural crops was lost in the 20th century.

Hemp

Hemp was the world's largest crop for 3,000 years as a source of paper, clothing, fuel and food. The American constitution was written on hemp. The Founders grew hemp. It can be cultivated without the use of chemicals or pesticides. If the government required all paper to be made from hemp the world's threatened forests could be saved. But it has gotten mixed up in narcotics laws and big-business interests.

Cotton, hemp's substitute, is responsible for more than half the chemicals used in US agriculture. Cotton subsidies are destroying the livelihood of farmers in Africa.

A Citizens' Jury

the locals know what aid they need

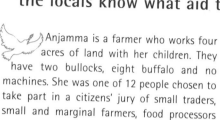

Anjamma is a farmer who works four acres of land with her children. They have two bullocks, eight buffalo and no machines. She was one of 12 people chosen to take part in a citizens' jury of small traders, small and marginal farmers, food processors and consumers to assess ambitious plans for farming in their part of India.

The Chief Minister of Andhra Pradesh, Chandra Babu Naidu, has been widely acclaimed for the state's modernizing achievements with Information Technology. If IT, why not agriculture? US agricultural advisers were brought in and Vision 2020 was drawn up, to bring millions of poor farmers straight into the 21st century with massive consolidation of farms, mechanization of agriculture, irrigation projects, new roads and the introduction of genetically modified crops. The state government claims that the program will eradicate poverty.

The Citizens' Jury sat through days of evidence from politicians, seed companies, academics, aid donors and NGOs in June 2001. Three scenarios were presented and advocates for each tried to persuade the jury that their particular approach would provide the best opportunity to enhance livelihoods and food security in the coming twenty years. An oversight panel checked that each proposal was presented in a fair and unprejudiced way.

The first proposal was the State's Vision 2020. The second was for environmentally-friendly farming to meet western demands for organic and fair-trade produce. The third was based on increased self-reliance for rural communities, low input, local food production and local marketing. The jury could either choose one of the pre-formed visions or promote its own unique view. Its verdict was very clear:

- They wanted encouragement for self-reliance and community control over resources.
- They gave priority to the maintenance of healthy soils (the farmers among them were conscious of the harmful legacy of pesticides and fertilizers).
- They wanted to maintain diverse crops, trees and livestock.
- They wanted to build on indigenous knowledge, skills and local institutions.
- They wanted to maintain control of medicinal plants and their export.

And their opposition was clearly expressed:

- They were horrified that Vision 2020's proposals would reduce those working the land from 70 to 40% in Andhra Pradesh. This would uproot 20 million people, leaving them with no livelihood.
- They did *not* want genetically modified crops, specifically not "Vitamin A" rice, nor "Bt cotton."
- They did *not* want consolidation of ownership into fewer and bigger farms for cash crops.
- They did *not* want an increase in contract farming.
- They did *not* want labor-displacing machinery.

Anjamma was asked what she would do if Vision 2020 goes ahead. "There will be nothing for us to do," she replied, "other than to drink pesticides and die."

In the US the Jefferson Center has conducted citizens' juries since 1974 covering a wide variety of issues like property tax reform, assisted suicide, environmental risks, traffic congestion, pricing, agriculture and water policy, electricity futures and many more.

> Citizens seem more intelligent than governments

We all believe in democracy but we all know that decision-makers are swayed by lobbyists or activists. The Citizens Jury provides an unparalleled opportunity for a group of citizens to learn about an issue from experts, and for legislators to hear the people's authentic voice. It has worked well in such diverse countries as the US, Germany and India, and skeptics have been surprised at the positive and sensible outcome of these juries.

www.jefferson-center.org

POPs & EDCs

synthetic time bombs

In January 1999, in Belgium, two gallons of transformer oil were accidentally mixed with fats destined for cattle feed. The oil contained two synthetic compounds, PCBs and furans. Ten feed manufacturers sold their products to 1,700 farmers, who gave them to countless chickens, pigs, and cows.

Although the exposure in each mouthful of feed was minuscule, the contaminants concentrate in animal fat. By the end of May, retailers throughout Belgium dumped all poultry and egg products, farms slaughtered any animals suspected of carrying the poisons and governments throughout the world banned importation of all Belgian animal products.

PCBs and furans are POPs - Persistent Organic Pollutants, and POPs are a legacy of the agrochemical revolution. They revolutionized farming. They kill pests and increase food production. The benefits were so widespread that anyone questioning their use was accused of being anti-science, obstructing agricultural and industrial progress and depriving the poor of protection against disease – exactly what politicians are saying about genetic engineering today. The National Academy of Sciences now says that POPs have caused severe damage around the world. Many are banned.

We should not be surprised. These chemicals were developed for explosives and nerve gases during World War II. At the end of the war the chemical companies were left with huge overcapacity, so they lobbied. And they succeeded. **They diverted their weapons against people to weapons against the land.** Not surprisingly, fifty years of chemical attack have degraded the soil and damaged the nation's health.

POPs are new to nature. They have not formed part of life's evolutionary process so they offend the first principle of sustainability. They are now found everywhere, even in the Antarctic. And they are working their way up the food chain from gulls and fish to seals and eagles. They accumulate so that we all now have 500 synthetic chemicals in our bodies. Mothers have many, including dioxin and PCBs, in their breast milk. POPs cause cancer and cancer's incidence is steadily rising, along with a massive cure (but not prevention) industry. It is not possible to dispose of POPs. We can only shift them from place to place. Bacteria won't degrade them. They don't disappear when attached to biodegradable diapers. They can't be neutralized from incinerator smoke stacks. Having killed someone, they are free to roam again. **POPs are in the environment and they will stay in the environment – forever.**

Five of the 12 most common POPs are also endocrine-disrupting chemicals (EDCs). The endocrine gland controls hormones, the central nervous system and other key functions of the body. Products can be tested at a few parts per billion for carcinogenic properties, but a few parts per trillion are capable of affecting hormones. Strangely, relatively large doses of EDCs might have no effect, but when tiny doses are administered to rats at a key stage of pregnancy, such as the day on which their hormones trigger limbs to develop, they have a devastating effect.

This may explain some strange findings. In the St Lawrence River, which is heavily polluted, the population of Beluga whales continued to decline after hunting stopped. It was found that their testes and ovaries were horribly deformed. Similar problems have been found with alligators, polar bears, monkeys, gulls, otters and shellfish. Feminized fish populations are found in nearly every river.

The chemical industry denies that POPs or EDCs can harm humans ("where are the corpses?") and they put massive pressure on politicians to allow them to continue unchecked with their hugely profitable, destructive industry. But there is a steady increase in autism, attention deficit hyperactive disorder (ADHD), dyslexia and uncontrollable aggression, which now affect one in five American children. Thousands of babies in Puerto Rico are experiencing premature breast development. The sperm count of men, globally, has reduced by 50% in the last half century. These symptoms, and many more, are directly related to disruption of the endocrine gland. Genetically modified crops are not tested for endocrine disruption, though there is more reason to suspect that changed genes will be harmful than changed chemicals. The chemical and agrochemical industries are conducting a gigantic uncontrolled experiment on humankind.

Pharmaceuticals also. "Pharming" these drugs as crops will spread pollen to other crops. Traces of contraceptives and many other synthetic drugs are already found in plants, mammals, birds, fish and drinking water. We will all be ingesting drugs prescribed for others.

Synthetic time bombs are ticking. The United Nations, with "unprecedented urgency," is trying to understand the problem and wishes to phase out the use of POPs altogether, against strong opposition from industry and from the US. Sweden is taking the only logical course – in four years it will be illegal to make or use persistent synthetic chemicals, whether or not they are known to cause harm. The WTO, of course, considers this a barrier to trade and may force these poisons on Sweden's citizens.

POPs are also the most important ingredients in the vast modern chemical industry that has more than 75,000 synthetic chemicals on the market. Nothing is known about a third of those in common use, only 10% have been tested for carcinogenic properties and hardly any for hormone-disruption. It is only logical that the industry should have to prove that its chemicals do not cause harm, but it has so much money and its lobby is so strong, that it is the consumer that has to prove harm – assuming he or she is not already incapacitated or dead.

First generation immigrants to the US have the low-cancer incidence of their native land. Third generation immigrants have the high-cancer incidence of the US.

Our Stolen Future
Theo Colborn et al 1996

Antibiotics

a gift of nature rejected

In just 60 years we have thrown away the advantages of a modern medical miracle.

The first of the five "kingdoms of life" on earth consisted of bacteria. Some of these are also known as germs because they can invade the human body and cause disease. Bacterial infection, however, can be treated by antibiotics, and the first, penicillin, became available in the 1940s – a medical miracle. There are now 15 types of antibiotic, though more may be developed, each of which acts in a different way.

High doses of antibiotic kill bacteria. But when bacteria are subject to low-level doses, some survive and some die. The ones that survive have a reproductive advantage. By this process the bacteria quickly develop immunity. This is why doctors only prescribe antibiotics when necessary and insist that you complete the course.

Modern surgery was not possible until 1945 when penicillin could kill the most infectious agent, *Staphylococcus Aureus* (SA). By 1947 some SA had become resistant to penicillin. In 1957 methicillin was introduced and methicillin-resistant-SA bacteria (MRSA) developed within two years, making the wards of some hospitals unusable. Now only Vancomysin controls MRSA's worst excesses, but for how long? Vancomysin-resistant bugs have already been detected. Within a few years SA may again become untreatable.

Like humans, animals that are kept in unnatural and crowded conditions are particularly susceptible to disease. So what do farmers do? They feed their animals regular doses of antibiotics! The appalling condition of battery chickens is well known, but systematic cruelty and overcrowding are common with many forms of animal husbandry. So we should not be surprised that resistant bacteria develop in these conditions and go on to affect humans; one's only surprise is that the experts are surprised. It has taken till now for Britain's government to admit: "We believe that the evidence shows conclusively that **giving antibiotics to animals results in the emergence of some resistant bacteria which infect humans."** Fish farming also uses antibiotic prophylactics.

What sort of society are we that it is only fear for our own health that has made us take any interest in providing healthy and humane conditions for animals?

You may think that plants do not carry antibiotics, but they do now. Most genetically modified (GM) crops, which have been planted all over America, carry antibiotic-resistance marker genes. Bacteria must be jumping for joy because this low-level presence will allow them to pick up the resistance genes.

Why has there has been so little control over the use of antibiotics in agriculture? It is only a question of time before superbugs will result from the routine use of antibiotic prophylactics in animal feed and from the implantation of antibiotic resistance marker genes in GM crops. When this happens we face the prospect of untreatable epidemics sweeping the world. Will the first pandemic arise from chicken batteries or from the GM plains of mid-west America?

- 60% of infections picked up in hospitals are now drug-resistant. The World Health Organization (WHO) says that it is only a question of time before microbes become resistant to the last effective antibiotic.

- The British Medical Association has called for antibiotic-resistance marker genes to be banned – rather late in the day. They are in GM crops all over the US and on test sites in Europe.

Against the Grain
Marc Lappé and Britt Bailey 1999

Microbes

tougher than humans

Human beings may be clever, but the humble microbe will outlive us.

Let's start with the big picture. It took four billion years for living cells to transform the virgin soup of the atmosphere – a toxic, chaotic mixture of methane, sulfurous compounds, carbon dioxide, and other substances – into conditions that could support life. We must thank the microbes for creating quality out of chaos. In the last few decades we have started to reverse all this. We are digging up everything the microbes "fixed"; we are creating novel materials and gases and disgorging them into the sky and onto the earth's crust. The earth is now running backwards into the chaotic garbage heap from which it started.

But the microbes are still at work. Bacteria, viruses, protozoa, yeasts and algae live in the soil, in water, in the air, on our skin and in our bodies, and are key players in maintaining suitable conditions for life. They recycle nutrients for plants, and in our gut they provide vitamins and aid digestion. Though we can't see them, we use them to make wine, bread, cheese and yogurt.

Microbes have awesome power. A single bacterium can, given the right conditions, take less than a day to multiply to a thousand billion cells. So a **policy of understanding and cooperation, as practiced in organic farming and in much eastern medicine, might be more enlightened** than zapping them whenever they harm our crops or cause us pain.

Our close contact with domestic animals at the dawn of civilization allowed microbes to jump from them to us, originating many of our "crowd" diseases: smallpox and tuberculosis came from cattle, the common cold from horses. We developed partial immunity through long contact, but Polynesians and original Americans had virtually no defense when they first came into contact with Europeans carrying these diseases. Pigs have had a bad record: they were responsible for the Asian flu epidemic of 1918-20 which killed 20 million people, more than died in the Great War; and there have been more recent incidents of viruses originating from pigs, particularly in Malaysia. This suggests caution over xenotransplantation, where pigs are used for growing human spare parts.

Other viruses may spread from forests where, until recently, humans have had little contact with wild animals. AIDS spread from Zaire in the 1960s but why in modern times and not before? Some scientists think that the AIDS epidemic might have started when a million Africans were used to test a new polio vaccine derived from chimpanzee kidneys in 1957-60, and they urge caution when handling unfamiliar or novel viruses. We are now coming into close contact with a host of unfamiliar microbes as virgin forests are being felled; infections are expected to jump when their natural hosts deplete.

In the 1960s many doctors believed that infectious diseases would soon be conquered, but this hope is now fading and **medicine is increasingly seen as a race to keep ahead of pathogenic microbes.** As soon as cures or new drugs are found, resistant bacteria seem to develop. Many hospitals have had to close wards due to untreatable bacteria or viruses, particularly MRSA. Malaria is spreading north from the tropics. Tuberculosis, which was thought to have been beaten, is coming back, and some strains are already untreatable so that, in New York, the medieval practice of forcible isolation has been reintroduced. People feel vulnerable to diseases that seem to be out of control so new infections like West Nile virus can cause panic. But, when everything in sight is sprayed, the spray may cause more damage than the infection it is supposed to control.

Human presence is the flicker of an eyelid in the timetable of microbes. We cannot assume that humans are somehow above the evolutionary battle and are bound to survive whatever damage we do to other forms of life on earth. A little less arrogance would serve us well.

Genes

and genetic experimentation

Genes are units of heredity – they reproduce themselves from one generation to another. They are fragments of the DNA sequence that have cohesive ends enabling them to be extracted and recombined, hence genetic experimentation is often referred to as "recombinant DNA" technology.

Forty years ago, molecular scientists dreamed up the wonderfully simplistic "central dogma" - that genes fully account for inherited traits. This gave rise to a multi-billion dollar industry that has made a massive assault on American agriculture. And the Human Genome project, that found the DNA sequences of all the genes in all 23 pairs of chromosomes that exist in every human cell, was heralded as "the ultimate description of life."

Instead, the Human Genome project destroyed the scientific basis of genetic engineering. It found that there are far too few genes to account for the complexity of our inherited traits, fewer even than possessed by a blade of grass. An industry that claims to lift agriculture onto new levels of scientific precision is actually based on a science and a technology in their infancy, at the stage of dropping toys out of the cot to see if they break.

Unfortunately, hype reached the highest political level, and the height of absurdity, with President Clinton describing the genome as "the language in which God created life." Few noticed the sacrilege of suggesting that God could only speak one language.

The human body is made up of cells. An adult has about a hundred thousand billion cells and most cells have a complete set of the genes. The genes turn themselves on or off in different ways in different cells according to their environment in what has been described as a symphony of gene expression. This may explain why chimpanzees, which share 98.5% of our genes, are different from us in almost every detail. A single gene can give rise to multiple proteins, so single function genes are a myth. Scientists are undecided whether a genetic disorder is due to the malfunction of a gene, a protein, a chromosome or a cell. There are fewer genes than different proteins in the body and some proteins, called prions, can reproduce themselves without involving DNA at all. We should be willing to accept how little we truly understand about the secrets of the cell, the fundamental unit of life.

Biotech companies are playing with just one aspect of life that has interconnections at all levels, and the genetically engineered crops now being grown represent a massive uncontrolled experiment whose outcome is inherently unpredictable. There must be great dangers in overriding patterns of inheritance that are embedded in the natural world through the process of evolution.

Some dangers were appreciated at an early stage. James Watson, joint discoverer of the structure of DNA, issued, with other scientists, an alarming open letter in 1974. Genetic experiments will "result in the creation of novel types of infectious DNA elements," they said "whose biological properties cannot be completely predicted in advance... new DNA elements introduced into E.coli might possibly become widely disseminated among human, bacterial, plant, or animal populations with unpredictable effects." They called for a moratorium on high-risk recombinant DNA experiments. But within a year the potential profits from genetic experimentation became irresistible and many scientists were hooked. In Scotland and Japan, as predicted, there have been incidents where modified E.coli has escaped the laboratory with disastrous results. Dr Mae-Wan Ho suggests that this technology could be the end of humanity. Dare we ignore her?

Changing the genes in every cell of an organism is massively complex. But to modify the egg and sperm before genes divide is relatively simple because you only change a single cell. This is called "germ-line therapy." The modified gene will reproduce itself in every cell of the organism as it grows and will be present in future generations. Germ-line therapy can therefore be described as artificial evolution. Whether the gene will stay in the same location in the DNA of second and third generations is disputed and other changes that might occur are not known.

We should question whether we will ever fully understand life and whether it is right for the science to be driven by commercial incentive. History is full of our mistakes. Mistakes with life will have a life of their own.

If the science can avoid too much damage through its early hubris, the manipulation of genes under strictly controlled laboratory conditions may have great benefits. The development of insulin and skin grafts are obvious examples and our genetic *understanding* could greatly help sustainable organic farming techniques.

E.coli is normally harmless. The deadly E.coli 0157:H7 first appeared in the US in 1982 and 9,000 people were affected in Japan in 1992. Last year there were 1,084 cases in the UK.

Some laboratory rats were exposed to radiation that gave rise to birth defects. The radiation had not changed the genetic code of the offspring. Then the fourth generation of the rats was also found to have the same rate of birth defects and mutations as the original parents. The researchers could not understand why the genome had become prone to mutations since the inherited genetic code had not changed. Radiation had destabilized the entire genome of later generations, not just the irradiated sections of the original rat. This indicated that unknown factors come into play with inheritance over several generations that cannot be explained simply by genes.

Genetic Engineering
Dr Mae-Wan Ho 1998

Making New Plants

improving on Creation

There is a popular perception that genetic engineering (GE, GM, GMO) is just a development of plant breeding techniques that have been practiced since the dawn of civilization. No honest scientist would make this claim.

Conventional breeding transfers genetic information between related organisms – members of the same species or, rarely, of closely related genera. Genetic engineering, however, overcomes the barriers that have existed for three billion years of evolution and now transfers genes between unrelated species, genera or kingdoms. For example, a potato may now carry the gene that stops a fish freezing.

Gene transfers have a planned result, but the process relies on trial and error. Occasionally the transferred gene seems to have the desired effect in the host organism but this may change during the plant's life or in future generations. Several processes are involved in the transfer:

• A powerful **vector** must be used, usually *Agrobacterium tumefaciens,* a bacterium that causes tumors in plants by inserting DNA from its own genetic code.

• A **promoter** is necessary to overcome a plant's natural resistance to infection. Usually the Cauliflower Mosaic Virus (CMV) is used, a virus that can cause damage to the pancreas, liver and brain of animals that ingest it. It "turns on" the transferred gene (hyperexpression) but often turns on other genes as well.

• A **marker gene** is added because genetic modification is a haphazard process and the scientist needs to know where the gene has arrived in the host genome. Until recently antibiotic-resistance marker genes were usually used. It has now been found that these genes can be taken up by bacteria in the gut. The British Medical Association says that the use of such genes is "completely unacceptable."

• **Random extra genes** get transferred. These are not planned, but strands of unexpected DNA have been found, for example, in Monsanto soybeans. The inserted DNA may also be unstable in both the initial plant and in future generations.

These are some of the ways in which genetic engineering alters plants. Is this really just an extension of conventional breeding techniques? The biotech industry, with strange logic, claims that genetic engineering is no different from normal plant breeding yet it is a radical new science. US agencies define GM food as "substantially equivalent" to ordinary food, so people do not need to know which they are eating.

But Europe was worried that independent testing of GM foods had not been carried out and it imposed a five-year ban on growing GM crops. European regulations have cost US farmers $12 billion in lost trade and the US has complained to the WTO that this is an infringement of free trade rules because damage could not be proved. European concern has now spread back to the US.

Allergies are a major problem. The National Academy of Sciences warns that genetically engineered plants may introduce allergens into pollen, which then spreads them in the environment.

Biotech companies insert toxins into every cell of some plants, our food, as insecticides. The toxins are also released into the ground where they kill susceptible insect larvae and degrade the soil. Low-level toxins in the entire cellular make-up of crops provide ideal conditions for resistance to develop in weeds, insects and bacteria. So dangerous compounds like paraquat and atrazine are being re-introduced to outsmart the pests, though they in turn will become ineffective. In most GM crops the use of herbicides and pesticides is increasing, often just before harvest, so the amount of residues (POPs and EDCs) left on food is also increasing.

Now for the really scary stuff: "horizontal gene transfer." Genes are units of heredity – part of the magic that shapes future generations. They go "vertically" from plant to seed, parent to baby. But Harvard University has found that modified genes also move "horizontally" from plants to bacteria. They can move from food to the bacteria in our gut and breach the lining of the womb; we are warned that baby food should not contain GMOs, and it is wise for an expectant mother to avoid GM foods. Most Americans want food with GM content to be labeled but biotech companies have, so far, prevented it.

GM foods have been consumed for seven years without harm, they say. But during this period food-derived illnesses have doubled. In California the number of children with autism has doubled in the last four years. The sperm count of men is steadily reducing. Allergies have greatly increased. But no one knows if this is consequence or coincidence because no monitoring of the long-term clinical or biochemical effects of GM has been done.

There are regulations to protect food for humans, but how effective are they? StarLink corn was approved only for cattle in 1997 because it contained a protein deemed to be allergenic for humans. But it escaped from animal feed and has been identified in food in countries from Canada to Japan. "My eyes were closing. My lips were numb and swollen. I felt like I had been knocked around by Mike Tyson" said one victim. StarLink contained Cry9C. A study sponsored by the EPA then suggested that the closely related Cry1Ab & c, used in *Bt* corn, could elicit antibody responses consistent with allergic reactions found in some farm workers. Some scientists then called for independent toxicological tests. Until this is done the allergenicity of GM corn is not known.

In 2000, researchers from Berkeley found GM contamination in native Mexican corn as well as indications that transgenic DNA was not stable. The significance of this was demonstrated by the ferocity of the biotech industry's attacks, not on the research, which had been reproduced and confirmed, but on the researchers. Publication endangered a $50 million grant from Novartis to their university, so the researchers also earned the hostility of their fellow scientists.

Does GM food affect our health? Little independent research on health effects of GM has been done in the US. In Britain, the first such research was so badly handled that it shattered public confidence. The unease increased as details emerged. According to four witnesses at the Rowett Institute in Scotland, telephone calls: first from Monsanto to President Clinton, then from Clinton to Tony Blair, then from Blair to the Institute - stopped this key project.

Dr Arpad Pusztai, one of the world's leading experts on plant lectins with 35 years of lab experience, in a project won in competition over 28 other teams, had found that GM potatoes with snowdrop lectin expression caused rats' brain, liver and heart size to reduce and also weakened their immune system. "I believe that this technology can be made to work for us," he said in a short interview on television in 1998 but "it is unfair to use our fellow citizens as guinea pigs; we have to find guinea pigs in the laboratory." The head of the Institute commented on "how well Arpad had answered the questions" and "a range, of carefully controlled studies underlie the basis of Dr Pusztai's concerns." The next day Pusztai was told to hand over his data, all GM work was stopped, the team was dispersed and he was threatened with legal action if he spoke to the press. He was banned from his laboratory and his phone calls and emails were diverted. Later, at the government's request, the Royal Society, Britain's leading scientific body, rushed out a condemnation of his work without obtaining the final report. Then his wife was fired from the Institute and his data were stolen from both his home and the Institute (what burglar would be interested in these?) However his work had been peer reviewed six times before being published in The Lancet (whose editor was then threatened), well received at a scientific conference and backed by 30 international scientists.

When research is contested, the scientific method is to reproduce it. In this instance, attempts were made to rubbish the 1995-98 study but not to repeat it. The public became acutely aware of the difference between science and commercial science. And confidence in the integrity of GM research plummeted.

Don't Worry
Andrew Rowell 2003

- Since GM crops were introduced in the US, farmers have been using more pesticides and herbicides, not less.

- Atrazine is a horror pesticide. It gets into rivers and aquifers, and disrupts hormones (an EDC). GM Liberty Maize allowed a farmer to stop using it. But weeds have already developed resistance and GM Liberty Atz is now used – corn with atrazine in every cell.

- Conventional crops have had a devastating effect on wildlife, as acknowledged by the National Academy of Sciences, but a three-year field trial study in Britain found that GM crops have an even worse effect.

- In 2003 the British government asked the public if it wanted GM food. 80% said "no." Over half said never. It also set up focus groups, citizens' juries, of randomly selected people with no allegiances. As they got to know more about it their opinion changed from neutral to anti; in particular that "no one knows enough about the long-term effects on human health."

- Percy Schmeiser had bred oilseed rape on his Saskatchewan farm for 50 years and was looking forward to retirement. He noticed some GM canola "volunteers" (left from previous crops) from a neighbor near his fields. Monsanto had also noticed. Most of his crops were not affected but 60% in one part were contaminated. He had never bought Monsanto's GM seed. Nevertheless, Monsanto served a lawsuit claiming that he had infringed their patent.

Judge W Andrew MacKay's ruling sent shock waves through the farming community, ruling: "the source of RR Canola is really not significant for the issue of infringement." Schmeiser lost $600,000 and contamination had ruined his life's work. "If I went down to Monsanto's headquarters and destroyed some of their plants by cross-pollination I would be thrown into jail." he said "Why does Monsanto have such a right?"

"This technology is totally different from traditional breeding techniques. Current laws were not written with this technology in mind."
Dennis Kucinich 2002

The Terminator

corporate control of food for profit

"The earth is not dying, it is being killed. And those that are killing it have names and addresses."
Utah Phillips

The people we are talking about in this chapter are probably decent family men; they probably attend religious services and believe in a loving God; they probably follow normal business practice. So why are they feared?

"Terminator" technology was a logical business solution. Rather than paying lawyers to sue farmers for saving and sowing patented seed, the seed could be genetically designed to commit suicide. This reduced its yield slightly, and did not exactly improve humanity's ability to cultivate, but Monsanto's friends in the United States Department of Agriculture (USDA) were happy to help. Let's just quote them.

Monsanto: "Terminator technology will open significant world-wide seed markets to the sale of transgenic technology, for crops in which seed currently is saved and used in subsequent plantings."

United States government spokesman, Willard Phelps: "Terminator technology's primary function is to increase the value of proprietary seed owned by US seed companies and to open up new markets in second and third world countries."

Melvin J Oliver, USDA scientist, not a company man, explains: "Our mission is to protect US agriculture and to make us competitive in the face of foreign competition. Without this there is no way of protecting the patented seed technology."

After intense worldwide hostility, Monsanto agreed to delay marketing terminator seed for further studies on environmental, economic and social effects. They grudgingly admitted, "We need some level of public acceptance to do our business." However, terminator seeds continue to be patented, including one that becomes sterile only after three generations – a farmer can be fooled into thinking he can replant the seed.

Other companies were also turning plant welfare on its head for commercial gain.

- AstraZeneca was developing seed that is sterile unless their own chemicals are applied.

- Novartis was even developing plants whose resistance to viruses and bacteria have been removed.

Plants that are sterile, plants that die without chemicals, plants that have no resistance to disease – these are among the achievements of biotechnology companies. And they claim that patents on life forms are necessary to enable them to continue to invest in this research and to prevent farmers in poor countries benefiting from their advances.

But that is not all; controlling humanity's food source is only a part of Monsanto's ambitions. To quote Bob Shapiro when Chief Executive, "It is truly easy to make a great deal of money dealing with primary needs: food, shelter, clothing."

"What you are seeing," said Robert Farley of Monsanto in 1998 after describing its purchase of seed companies around the world, "is not just a consolidation of seed companies, it's really a consolidation of the entire food chain. Since water is as central to food production as seed, Monsanto is now trying to establish its control over water. Monsanto plans to launch a new water business, starting with India and Mexico, since both these countries are facing water shortages. These are the markets that are most relevant to us as a life science company committed to delivering food, health and hope to the world, and in which there are opportunities to create business value."

Monsanto Strategy Paper: "We are enthusiastic about the potential of partnering the World Bank in joint venture projects in developing markets. The Bank is eager to work with Monsanto." Many Indians understand this to mean that Monsanto aims to control the vital resources of the Indian sub-continent, using public finances (the World Bank) to underwrite the investment. Indian agriculture would then be at the mercy of a private company motivated by profit.

Agricultural biotechnology in the hands of corporations is about control, a branch of empire.

- By 2025 the need for water in India will be 50% more than is available. The crisis will be even greater in 35 years time when the Himalayan glaciers, which supply summer water to the Indus and the Ganges, are gone. The water table in most states in India is dropping at the rate of one yard a year. Will water companies be more interested in thirsty peasants than industry?

- One company, Cargill, controls 80% of global grain distribution. The top ten corporations control: 85% of all pesticides, 60% of all veterinary medicine, 35% of all pharmaceuticals, 32% of all commercial seed.

- The government of Brazil, responding to the will of its people, banned the planting of GM Soya. Monsanto spent $600m in buying Brazilian seed companies and encouraging farmers to smuggle large quantities of GM seed across the border. Monsanto has not been charging them royalties. The government was forced to cave in and Brazil is no longer a GM free country.

- Monsanto's New Leaf potato incorporates pesticides in every cell and the potato itself has to be registered as a pesticide.

- In 2002 the massive failure of Bt cotton in India's southern states led to a ban on planting in northern states.

- AstraZeneca makes tamoxofen, a breast cancer drug. It also makes acetochlor, a pesticide identified as a cause of cancer. This forms a neat business cycle, perhaps suggested by its accountant.

The Monsanto Files
The Ecologist Vol.28 No5 1998

Nature in Balance

survival of the fittest?

The skeletal structure of the Oxford Museum, with a giant dinosaur across its main hall, is where the first debates on the origin of species took place. In a cabinet there are two bean-weevils that are so small they can sit comfortably on a match head but their rate of reproduction is prodigious.

Beside them is a glass jar. Dr George McGavin, the curator, says that it would take only about 70 days of unchecked mating of the weevils and their progeny for the jar to be filled with their eggs. If they continued thus they would fill the room in 100 days and the entire internal volume of the earth in a little over 15 months. If this happened they would, of course, have nothing to eat. By winning, they would be on the losing side. Nature, during four billion years, has evolved plenty of checks to stop this happening. If any species is too successful its population may suddenly collapse.

Each species needs a balanced habitat to provide it with food and each species is subject to a balance of predators and microbes. But the balance can be rocked. When a plant or creature from one part of the world is introduced into another, it may no longer have the natural restraints that had evolved in its original habitat. Biologists refer to the 10/10 principle: one in 10 alien organisms will thrive in a new environment and one in 10 of those that thrive will become a pest.

There were no rabbits in Australia until a few were deliberately introduced in 1859; they soon spread throughout the continent with disastrous results.

A single Japanese Knotweed, planted in a garden near Brighton, escaped into a rubbish tip and has now spread throughout the UK; it is unaffected by pesticides and Swansea has a full-time Knotweed officer who reckons it would cost $15 million to eradicate it from the city (it would return, of course). Mink escaping from farms have threatened the survival of some native species. The gray squirrel has displaced the red squirrel in many countries. The zebra mussel has caused havoc in the Great Lakes of North America. The comb jellyfish has colonized the Black Sea. Water hyacinth has clogged waterways in Africa and India.

Whether the 10/10 principle will apply to genetically modified plants and animals is an open question. Some novel organisms are being deliberately released as trial crops. In laboratories, novel bacteria and viruses are developed as vectors to transfer genes from one species to another. Genes from animals are being introduced into agriculture and human surgery.

But once novel plants, bacteria, viruses or animals are out in the open they have a life of their own and they are no longer in our control. As they are new to nature it is unlikely they will be subject to the constraints that have developed over millions of years.

Seen from space the biosphere seems a fragile skin, as delicate as the bloom on a peach. But humanity is infinitely more fragile. We do not control nature, nature supports us. Nature has evolved in a way that retains a balance. The forests act as the lungs of the world and contain unexplored genetic richness; plants and plankton begin the food chain on land and sea; micro-organisms make the soil fertile and enable creatures to digest food – but it is presumptuous comment on this infinite complexity. Creatures that disturb this balance find themselves rejected.

Where Next
Duncan Poore 2000

Population

more or less

The world population in 2050 will be decided by three billion women. If every second woman decides to have three rather than two children, the population will rise to 27 billion. The best guess is that, on average, women will have 2.1 children, the "fertility rate," and the population will stabilize. If, however, every second woman decides to have only one child instead of two, the world population will sink to 3.6 billion. Small changes in people's motivation result in huge changes to numbers.

Corporations use the scare of rising numbers to justify risky experiments with chemicals and genes, and politicians justify economic growth as necessary for controlling population.

Stable societies in the past may have understood "carrying capacity" – the ability of their environment to sustain the desired quality of life over a long period. For example, the early colonizers of America had larger families than

their relatives in crowded Europe. But improved agriculture and aid programs released families from this bondage and they felt confident in having more children. Better expectations and reduced infant mortality resulted in the population explosion. We face the paradox that the expectation of food security caused the world population to rise, but the number of hungry people is now greater than ever before.

Human population increased very gradually over thousands of years. Then, following the industrial revolution, the rate of growth suddenly shot up, particularly since 1950. Since the 1970s there has been a gradual reduction in the rate of growth in most parts of the world. The usual way of predicting numbers is to guess at fertility rates and assume that, since they have always increased in the past, they will continue to increase in future, but will level off in due course.

But demographers also use the technique of fitting a mathematical equation to past trends. This shows a startlingly different outcome. The graph levels off in 20 years' time and then drops rapidly to pre-industrial levels.

Falling confidence in the economy could provide feedback that reduces parents' perception of their future prospects, thus reducing the fertility rate. And global warming, spread of tropical diseases and increased conflict point to a higher mortality rate. These trends tend to confirm the mathematical approach. Today, more than 60 countries have fertility rates below replacement level and the rate of growth in other countries is reducing. As their populations get older, rich countries will compete for immigrants of working age from poor countries.

The surge of population during the last century may have been a temporary phenomenon brought on by access to the cheap energy on which almost all our technology depends. The graph of population numbers may end up resembling the bell shaped graph followed by the rise and depletion of oil production. In due course the human population may return to a level that is within the carrying capacity of the world.

Italy and Norway

It is ironic that Bianca in Catholic Italy has just 1.2 children, which signals a crash from 56 to 8 million Italians this century; whereas Astrid in Norway has 1.8 children. Why? Bianca is well educated and has prospects but there is poor state childcare provision and Italian men are not renowned for helping in the home. Astrid is just as keen to pursue a career but the government helps mothers to juggle family and work, about half the jobs held by women are part-time, crèches are universal, paid parental leave lasts for a year and her husband is better house-trained, so the population is stable.

World fertility rate:

1950 5.0 children per woman
Today 2.7 children per woman

Population Politics
Virginia Abernethy 1999

Patenting Life

wait a minute! Who made it?

Only "commerce" could think of allowing life forms to be privately owned. An African tribal-doctor might wish to keep his remedies secret but no one would deny others the chance to emulate him. Indians know trees by their medicinal properties, and the idea that anyone should pay a fee for using these properties is inconceivable. It was accepted that anyone in the world could use the resources of nature – until 20 years ago.

It happened in the US. A genetically engineered micro-organism, designed to consume oil spills, provided the test case. The Patents and Trademark Office (PTO) rightly rejected the application, arguing that living things could not be patented under US law. But at appeal the patent was allowed by a narrow majority on the basis that the microorganism was "more akin to inanimate chemical compositions than to horses and honeybees." The PTO, still believing that patenting life forms was wrong, appealed to the Supreme Court, which upheld the patent in 1980. But four of the nine judges strongly opposed it and it was nearly rejected.

Then in 1987 the PTO did an astonishing about-turn. It issued a ruling that all genetically engineered living organisms can be patented - including human genes, cell-lines, tissues, organs, and even genetically altered human embryos and fetuses. This ruling opened the floodgates.

Patented things have to be "novel, non-obvious and useful." No one has ever argued that oxygen or helium, for example, however non-obvious and however useful, can be patented just because a chemist has isolated, classified, and described their properties. These elements are not novel; they exist in nature. But with total lack of consistency the PTO now holds that anyone can claim a human invention simply by isolating and classifying a gene's properties and purposes, even though these are not novel but pre-exist in nature. No molecular biologist has ever created a gene, cell, tissue, organ, or organism *de novo*.

The first mammal to be patented was a mouse with human genes. The team that created Dolly the sheep applied for a broad patent to cover all cloned animals. Then, a businessman, John Moore, received hospital treatment for a rare form of cancer; later he found that the University of California, Los Angeles had patented his body parts and licensed them to a pharmaceutical company at a value of $3 billion; the California Supreme Court ruled that Moore had no rights over his own body tissues. A broad patent gives Briocyte worldwide ownership of all human blood cells from the umbilical cord of babies for pharmaceutical purposes, even though the company has only isolated and described the blood cells.

Broad patents have caused furor. Patenting of the properties of the Neem tree and Basmati rice by American companies has provoked violent anger throughout India. An American professor failed in his attempt to patent turmeric for medicinal purposes but it cost India $500,000 to fight the case. Papua New Guinea was furious to discover that the US government had patented cell lines from its citizens. The arrogance of this new colonialism is breathtaking. Patents take no account of the transformation of wild grasses, tubers, etc. into crops by indigenous people over millennia. They literally allow western companies to hijack knowledge that has been used by all – even nature itself – and charge others for using it. But poor nations fear sanctions if they do not adopt American practices on patenting.

The World took a wrong turn in America in 1987. Might India declare, unilaterally, that the fundamental elements of life are global commons and ought never to be up for sale to private interests at any price? All life knowledge, at present in private hands, would then be freely available on the sub-continent. Perhaps India would become the powerhouse of the life sciences because of this free availability of knowledge.

Commercial Eugenics

an alternative future for the poor

Economic inequality has become extreme. But redistribution of wealth is extremely unpopular among the rich who wield power. They may decide instead to solve humanity's biggest problem by taking inequality into new spheres.

Our mental capacity evolved to meet the needs of hunter-gatherers and is built into our makeup. Even isolated tribal people, given the right training, can use the Internet. So by imagining hunter-gatherer society, some believe, we can begin to understand some human oddities – man is promiscuous and woman coy because that helps his genes to multiply and hers to use their annual chance to obtain quality. It is a compulsive game that you play according to your interests.

A sociologist asks why we are able to co-operate well in groups up to 150 but are often hopeless at larger scale decision-making. An art historian asks why we like landscape paintings that have an edge-of-forest feel. Evolutionary psychology is a surprisingly static cause-and-effect type of theory.

Scientists have latched onto the popular cause-and-effect concept because it captures the imagination and helps to release funds. Almost daily, scientists claim to find genes for this and that: a gene for inherited disease, for alcoholism, for aggression, for depression, for dyslexia . . .

Finding and neutralizing the gene for Down's syndrome would be one of the great achievements of medicine. If air traffic controllers could be screened for genetic pre-disposition to black out, it would surely be our duty to protect the public. If we could find correlation between a gene and aggressive behavior perhaps we could reduce crime. In

California genetic scans are already used in the sentencing process to determine whether a convicted criminal is likely to re-offend.

Insurance companies are bound to ask for genetic screening or at least for us to declare any tests that show genetic pre-dispositions. Employers will want to check that their investment in trainees is unlikely to be wasted. The billionaire will want to ensure that his progeny are tall, intelligent and sociable – all traits which, it is claimed, can be enhanced by genetic modification. We are at an early stage of the science, but if these techniques become possible they are bound to be adopted.

Instead of having a genetic make-up that was developed for hunter-gatherers and is ill adapted to modern society, we can now change our genes to suit our present needs, they say. In our market-driven society this will no doubt be achieved through personal and commercial incentives.

Molecular Biologist Lee Silver of Princeton University regards himself as one of the scientists blazing a trail to a future containing the Gene-Rich and the Naturals. The Gene-Rich, offspring of today's super-rich, will be about 10% of the population, enhanced with synthetic genes and having the life-span of Methuselah. The Gene-Rich will become the rulers of society – businessmen, musicians, artists, athletes and scientists. "With the passage of time," Lee Silver predicts, "the genetic distance will become greater and greater, and there will be little movement up from the Natural to the Gene-Rich class. Naturals will work as low-paid service providers or as laborers" – no doubt with good television to divert them. "Gene-Rich humans and Natural humans will be entirely separate species with no ability to crossbreed and with as much romantic interest in each other as a current human would have for a chimpanzee."

The higher species may treat us as we treat lesser species. A humane cull may be necessary to tackle one of our biggest problems – the adverse environmental effects of excessive numbers.

The Biotech Century
Jeremy Rifkin 1998

Nanotechnology

success might be the worst-case scenario

Deconstruct a potato atom by atom, store the information on a computer, broadcast it to the other side of the world where they can feed dust into a few trillion nano-assemblers and, hey presto, they reconstruct the potato. That may be an extreme description of what could be the defining technology of the coming century.

Nanotechnology aims to manipulate atoms. At this scale air and rock, life and non-life, mind and matter, are all assemblies of atoms arranged in different ways. It is the ultimate "convergence" science that spans all academic disciplines from physics to biology to health to cognitive neuroscience. A lay person can't hope to understand the science, but it sounds exciting: "it will solve humanity's material needs," "we can all have a better quality of life without having to work," "electricity will be too cheap to meter," (we have heard that one before) "neurons could be re-engineered so that our minds talk directly to computers or artificial limbs," "fine tuning our metabolism could stop aging," "computer networks could be merged with biological networks to develop surveillance systems," "viruses could be re-engineered to act as weapons." These are some of the longer-term claims. A shorter-term prediction recorded by *The Economist* may be a more realistic reason for the hype: "It will be a trillion-dollar business in ten years time."

For 200 years new sciences have transformed human life and capabilities so we are always optimistic that the next science will solve the problems left by previous ones - and those who question any exciting new science are dismissed by politicians as Luddites. But a quick review suggests caution.

The Industrial Revolution has given developed countries undreamed of riches, but it also started the process of global warming. Nuclear science threatens global holocaust. Persistent synthetic chemicals have worked their way up the food chain and now affect reproduction and the central nervous system of animals and humans. Antibiotics have been a temporary miracle but now we again face pandemics. The jury on genetic science is still out, and it is controlled by corporations for profit.

So new sciences have both good and bad aspects, and the more powerful the science the more extreme these effects can be. Nanotechnology may be the most powerful technology that has ever existed, with wonderful potential. But ultra fine particles (UFPs), for example, of a size that is extremely rare in nature, can penetrate skin and alveolar membranes, leaving the body wide open to toxic effects – and UFPs are already being used in a wide range of applications from drug delivery to sun creams. A particularly jolly warning by none other than Bill Joy, managing director of Sun Microsystems, is that out-of-control replicating nanobots might reduce global ecosystems to dust in a matter of days; but, keep our fingers crossed, that may just be imaginative science fiction.

Sciences develop in phases: First excitement. Then hard sell. Then reality. Nanotechnology is at the beginning of the second phase.

Unfortunately the greatest possible success for the science could be the worst-case scenario for society. If the technology lives up to its promises, if human effort is replaced once and for all, if our life can be spent pursuing art, literature and video games, if danger has to be artificially created to give us thrills, then human nature changes. Proponents of nanotechnology talk cheerfully about a post-human future. But human meaning is produced by human effort and, like King Midas, we may be deprived of the life we need to live.

NBIC. "Nano-Bio-Info-Cogno" is a project that has received the biggest research grant since the space program. It will use nanoscience for brain implants, genetic modification, pharmaceuticals, nuclear weaponry, intelligent uniforms etc. in a strategic bid to extend US commercial and military dominance.

Understanding Nature

and re-designing life

Our understanding of the world around us is determined to a great extent by the culture in which we live.

St Thomas Aquinas saw the natural world as a Great Chain, a myriad of plants and animals in a descending hierarchy of importance. For him nature required dependent relationships and obligations among creatures that God had created. Diversity and inequality guaranteed the orderly working of the system. His portrayal of nature reflected the hierarchical structure of the medieval society in which he lived.

Later, Darwin's portrayal of nature also reflected the society in which he lived. The capitalist marketplace was competitive, where only the fittest survived and his theories about plants, animals or man reflected this. As with the perfecting of machines, evolution in nature produced better and better models. His theory precisely matched Adam Smith's "invisible hand" that enabled acts of individual selfishness to result in general well being. The intrinsic value of each living thing in the medieval paradigm was replaced by mere utility value. Individual, clan, village and town structures, in which people could choose how they wish to live and what they wish to protect, gave way to types, categories and the dehumanization of politics. This allowed dictators or political cliques to act out theories in which people – as machines or competing economic units – could be manipulated to suit a particular social theory. It led directly to communism, to the eugenic experiments of the last century, and to market fundamentalism.

Nature is now being cast in the image of the computer and information science. Creatures are no longer birds and bees, foxes and hens, but bundles of genetic information with no sacred boundaries between species. In commerce, new groups of companies are emerging with shared relationships within complex embedded networks that are able to respond quickly to fast-changing flows of information. A generation brought up in a computerized society can accept information as a basis for understanding nature. It reassures us that the infinite and chaotic complexities and multi-faceted expansions of all our activities, our ever-shifting life-styles, and our experimentation with genes, are in harmony with the evolutionary processes of nature. Thus we avoid feeling confused and threatened.

In the 1940s Norbert Wiener pioneered the concept that "all living things are really patterns that perpetuate themselves. A pattern is a message and may be transmitted as a message; the fact that we cannot telegraph the pattern of a man from one place to another seems to be due to technical difficulties." This is a concept that will be familiar to viewers of Star Trek.

The ability to handle information, rather than to develop knowledge or wisdom, is typical of modern life: today's understanding gives way to tomorrow's; we must be continually open to new scenarios; through drugs and plastic surgery we can reinvent ourselves; life and work are games to be played.

Molecular biologists no longer talk of laws of nature or objective reality, but of "scenarios," "models," "creative possibilities" – this is the language of architects or artists. Those at the cutting edge see their ability to handle almost unlimited information through computers (soon to be living DNA computers) as a natural extension of evolution. Evolution has now passed into their hands (they say "into humanity's hands" but who else has the ability except molecular biologists?) **Molecular biologists are the artists of new forms of life; they will modify and design future generations.** In other words they are the invisible hand of Darwin's natural selection or the God of St Thomas Aquinas.

But a new consciousness is emerging that approaches the world we share with other cultures, other creatures and a myriad of life forms, with awe and reverence.

Intuition

common sense, imagination and morality

Science is largely responsible for our progress and prosperity, and it is natural to look first to scientists for the answers to difficult questions.

But science, almost by definition, is reductionist - it looks at the simple constituents of complex things. Science can improve crop yields but not get the additional food to hungry mouths. Scientists have made no secret of their ignorance about the political effects of nuclear energy, and these effects have proved more important than the science itself. Governments like to claim scientific support partly because it allows them to make decisions within a cabal of experts, bureaucrats and corporate managers, hidden from the public behind obscure language. But recent events suggest that we should give more weight to intuition, imagination, common sense and morality.

Intuition might have prevented the British turning cows into cannibals – they would have avoided BSE, cattle would have suffered less, and they would have saved $8 billion.

Imagination could show that clear labeling of content and origin would enable epidemiologists to trace what is happening in a complex food system, and help people to make informed decisions on what they wish to buy.

Common sense would tell us that a centralized food system is dangerous – without such a system two gallons of transformer oil would not have destroyed Belgium's entire food economy in 1999. Common sense surely tells us that planting crops that are genetically new to nature, on the basis of "lets see what happens," could be rather like releasing rats to see whether or not they spread bubonic plague. Common sense tells us that the cost of coal, oil, gas and water should reflect depletion and contribution to climate change, not just the cost of extraction.

💡 **Morality** tells us not to keep chickens in such crowded conditions that they can't walk, live only six weeks and are then shackled upside down to a conveyor belt before being killed. Without this cruelty we would not have salmonella. Morality tells us to ban BST hormone treatment because it causes acute suffering to cows, let alone the suffering it might cause humans. Morality tells us not to participate in inhuman trading practices. Morality, eventually, told us to stop slavery.

When intuition, imagination, common sense or morality – let alone science – suggest that a new policy, product or procedure is suspect, it **should be the responsibility of the promoter to prove that the objections are unfounded.** This is the "precautionary principle." Our government and the World Trade Organization act on the basis that no restriction must be allowed until conclusive evidence shows harm. By then the damage is done.

Some countries use citizens' juries, chosen at random, who are presented with some technical background and a host of expert and lay views as to whether a certain area of research is worth pursuing. The citizens' jury then comes to a verdict as to whether the research seems reasonable and fair by criteria that they themselves develop. Both scientists and political theorists have been surprised and impressed by the results.

"Failing to understand the consequences of our inventions . . . seems to be a common fault of scientists and technologists. We are being propelled into the new century with no plan, no control and no brake."

Bill Joy
Chief scientist, Sun Microsystems

Two Japanese Farmers

more with less - a fascinating tale

Weeding rice paddy-fields is very labor intensive: so Japanese farmers eagerly adopted the chemical approach, even though it did not improve on the high yield of traditional Japanese methods. Only a few farmers held back from the Agrochemical Revolution.

The Furunos are a Japanese farming couple who do not use chemicals. Their system uses ducklings to fertilize the fields and eat the weeds. The rice has a very high yield and they get four different harvests: rice, fish (roach), eggs and ducks. In spite of a lifetime spent perfecting the system they are happy for it to be adopted by anyone. "Financial success is not important." They say, **"We did not patent the method, we just want it to be widely adopted."** Ten thousand farmers now use the method in Japan and it is spreading throughout South-East Asia.

Masanobu Fukuoka inherited a farm from his father. He claims to be lazy. "If things grow in the wild on their own, why does one have to do so much hard work?" he would ask. So in a short time he had destroyed his father's orange orchard and was having similar success with the rice. But he persisted with his unusual convictions and over the years developed a system of "do nothing" farming. He does not flood his rice fields, he does not weed them, he does not dig or plough and, above all, he does not use any chemicals. Yet his crops are resistant to pests, yield as much as traditional or chemical rice growing, and require much less work.

But the system requires careful management of crops and is related to a particular location. The farmer needs to know the microclimate and be observant. In the autumn he sows rice, clover and winter barley and covers them with straw. The barley grows immediately and is harvested in May and its straw scattered on the fields. The spring monsoon weakens the weeds and allows the rice to grow through the ground cover. Nothing more is done until the rice is harvested in October. Over the years the soil has got richer and richer with no plowing, without taking away the straw, and with no addition of manure or fertilizers.

"This is a balanced rice field ecosystem," he says "Insect and plant communities maintain a stable relationship here. It is not uncommon for a plant disease to sweep through the area, leaving the crops in these fields unaffected."

He has a similar approach to his restored orchard, harvesting a great variety of vegetables under the trees without cultivation but with a lot of care. And he has the time to be a philosopher.

Both farmers achieve yields that could make Japan self-sufficient in food, but the government persists in supporting chemical methods using alien seed.

"Because the world is moving with such furious energy in the opposite direction," Fukuoka-sen says, "it may appear that I have fallen behind the times, but I firmly believe that the path I have been following is the most sensible one."

The One Straw Revolution
Masanobu Fukuoka 1992

Mental Equipment

reality beyond our grasp

Our brains did not appear suddenly from nowhere; they evolved in the same way that other parts of our bodies evolved. Gradually, over hundreds of thousands of years before settled civilization, those with greater abilities supplanted others. So our mental equipment is appropriate for specific needs and is related to just five senses. It is not an abstract intelligence designed to comprehend all of existence, however much we use this equipment to understand and manipulate the world around us. It has limitations.

To appreciate the implications you only have to imagine a group that lacks one of the five senses, a group that has been blind from birth, discussing sight. They have heard people talk about color, about clouds, about mist, about distance and about beauty. How would they discuss these concepts among themselves? Some might say it is all a myth and deny that sight exists; some might be acutely frustrated that they can't quite grasp the idea; some might develop a form of words and bully others to accept their definition and others might be more tentative and try to feel their way into an understanding.

If our cognitive equipment evolved to meet the needs of hunter-gatherers, it is not surprising that some aspects of reality are simply beyond our grasp - just as a dog picks up only part of your meaning: "blah blah blah fido blah blah." But we do seem to sense further reality through a dark, distorting glass. Our vague perceptions are the stuff of mystics, poets, musicians and painters. Religions have tried to tie these realities down into concepts and words.

Even scientists describe things that fall beyond comprehension. Newtonian physics and Cartesian geometry made sense to our five senses. No longer. Matter is now known to be a mass of energy, more like a web of interpenetrating vibrations than like solid building blocks. Particles of energy are connected to each other in mysterious ways, seemingly unlimited by space and time. An experiment is changed by being viewed. Scientists now even suggest that "string," with ten dimensions and a whole world of magnitude below the size of molecules, could give clues for understanding the infinite vastness of an expanding universe. All this points to connections at a far deeper level than we will ever fully comprehend.

If there were creatures whose cognitive equipment had evolved beyond ours, they might be amused at the way we describe concepts that are mysterious to us but clear and obvious to them. But they might also be alarmed that our technology allows us to play with things whose full interconnectedness we can't possibly appreciate – they might see us wandering towards the precipice and be shouting **"for God's sake STOP!"**

I, the fiery light of divine wisdom,
I ignite the beauty of the plains,
I sparkle the waters.
I burn the sun and the moon and the stars,
With wisdom I order all rightly.
I adorn all the earth.
I am the breeze that nurtures all things green.
I am the rain coming from the dew
That causes the grasses to laugh
With the joy of life.
I call forth tears, the aroma of holy work.
I am the yearning for good.

Hildegard of Bingen
1098–1179

How the Mind Works
Stephen Pinker 1997

References

Organizations

Actionaid USA. 1112 16th Street NW Suite 540. Washington DC 20036: office@actionaidusa.org

Alliance for America: www.allianceforamerica.org

Alliance for Democracy. 760 Main Street. Waltham MA 02451: peoplesall@aol.com

ASPO The Association for the Study of Peak Oil: www.energiekrise.de. and: www.asponews.org

Best Foot Forward: http://bestfootforward.com

Center for Concern: www.coc.org. 1225 Otis Street NE. Washington DC 20017. alexander@igc.org

CNES Center for Network on Essential Services: www.servicesforall.org. 7000-B Carroll Avenue Suite 101. Takoma Park. MD 20912

Die Off, with links to articles: www.dieoff.org

Disinformation, alternative news and underground culture: www.disinfo.com

Earth Council: www.ecouncil.ac.cr

Earth Policy Institute: www.earth-policy.org

Environment Globe Post: www.environmentpost.com

FEASTA. The Foundation for the Economics of Sustainability: www.feasta.org

Footprint: www.iclei.org

Friends of the Earth: www.foe.org.

GCI: Global Commons Institute: www.gci.org.uk

Green Party-US: www.gp.org. PO Box 57065. Washington DC 20037

Greenpeace-USA: www.greenpeaceusa.org

INES, issues of global responsibility: www.inesglobal.org

The Jefferson Center, for citizens juries: www.jefferson-center.org. mail@jefferson-center.org

Land Value Tax Campaign: www.landvaluetax.org

Move On, democracy in action: www.moveon.org

Met office Hadley Centre: www.met-office.gov.uk

Mother Jones magazine: www.motherjones.com

Nature Research Journal: www.nature.com

New Internationalist US: www.newint.org

New Scientist-US MA: www.newscientist.com

Organics: www.ams.usda.gov/nop/indexIE.htm

ORG: Oxford Research Group, for peace studies www.oxfordresearchgroup.org.uk

Pew Center on climate change: www.pewclimate.org. 2101 Wilson Blvd., Suite 550, Arlington, VA 22201

Preparing for Peace: lecture series available on: www.preparingforpeace.org

The Progress Report: www.progress.org

Public Citizen: www.citizen.org. 1600 20th Street NW. Washington DC 20009

Rachel's Environment and Health Weekly: www.rachel.org

Rio + 10 (Johannesburg): www.rio-plus-10.org

Royal Commission on Environmental Pollution. 22nd report (UK official scientific research on climate change): www.rcep.org.uk

Scientific American: www.sciam.com

The Sky Trust, acknowledging that the sky is a global common. www.usskytrust.org.

Survival International. www.survival-international.org. For indigenous people.

The Natural Step: www.naturalstep.org. 116 New Montgomery Street, Suite 800. San Francisco CA 94105

Turning Point Project: www.turnpoint.org

UCS. Union of Concerned Scientists: www.ucsusa.org

UNDP. Human Development Reports (annual): www.undp.org

UNEP. Global Environment Outlook (GEO 2000 and annual): www.unep.org

US Global Change Research Program: www.gcrio.org

US NAS. National Academy of Sciences: www.nas.edu

US National Oceanic and Atmospheric Administration: www.noaa.gov

Vostok ice core data: http://cdiac.esd.ornl.gov

WTO, spin-free facts about global trade: www.gatt.org/trastat_e.html

WWF-US: www.worldwildlife.org

Yes magazine: www.yesmagazine.org

Z Communications, collected articles: www.zmag.org

Notes on some chapters

• **Introduction** • Full text of the Scientists' Warning: www.ucsusa.org. • "The Cultural Creatives" by Ray Anderson 2000.

• **Don't Predict** • The principles are derived from the four "System Conditions" of The Natural Step, but **see** his book for a full description. The conditions were developed by the oncologist Dr Karl-Henrik Robèrt and achieved consensus among scientists in Sweden. They provide a working definition for "upstream" action on sustainability. Most other definitions, such as Brundtland, have little practical application.

• **Ozone** • in 2002 there was an encouraging reduction of the ozone hole but in 2003 it was larger than ever.

• **Water** • for the disastrous effect of the current construction of the Narmada dam in India **see** "The Cost of Living" by Arundhati Roy.

• **Ecological Footprints** • The Ecological Footprint theory was developed by Wackernagel & Rees. • For a fuller discussion of Kerala see "The Growth Illusion" by Douthwaite and "Hope Human and Wild" by Bill McKibben • China warning is by Klaus Topfer of UNEP.

• **Atmospheric Carbon** • National Security Strategy Ch.6 p20 makes climate change a security issue. However the Bush administration has the most destructive environmental record of any and it has even censored the climate section of the EPA's annual scientific report • US NAS: "Abrupt Climate Change: Inevitable Surprises" 2001 • PEW Center: "warming in the US is expected to be higher than elsewhere ... sea-level rise will gradually inundate coastal areas,

increased risk of droughts and floods, threats to biodiversity and a number of potential challenges for public health. We must fundamentally transform the way we power our global economy." • For the Permian extinctions **see** "When Life Nearly Died" by Michael J Benton 2003 • **see** Die Off article on Climate Change.
• **Europe Cooling** • Potsdam Institute prepared computer model of Atlantic currents • Scottish Executive's Marine Lab. Aberdeen published measurements of reducing salinity "consistent with models showing the stopping of the pump and the conveyor belt" • Fisheries Lab of the Faroes records increasing temperature of pump water • Bergen University reported Greenland-Norway current into reverse.
• **Rio and Kyoto** • There is well funded denial of global warming by vested interests, originally led by the discredited Global Climate Coalition and now by Western Fuels Inc., an organization representing the coal industry.
• **Contraction & Convergence** • **see** also the Sky Trust, which proposes issue of gradually reducing number of carbon emission permits and dividing the income equally among all Americans. Its principles are: 1. The sky belongs to all of us equally. 2. The sky does not belong to corporations or the government. 3. Pollution must be limited to what the sky can safely absorb. 4. Once limits are set, companies should pay for pollution permits. 5. The money they pay should go into a trust. 6. The trust should pay equal dividends to all citizens. **See** chapter on Citizens' Dividend.
• **Future of Oil** • **see** ASPO newsletters for monthly independent updated assessment of oil depletion •

see also "The New Rulers of the World" by John Pilger 2000 • Seizure of Saudi oilfields and comments on Saudi investments in the US, **see** Report to the Pentagon by the Rank think-tank • **see** Die Off article on Oil Depletion • "Before the Wells Run Dry" by Feasta 2003
• **Nuclear Power** • DU **see** UN submission on Human Rights 1996.
• **Weapons of War** • **see** "Resource Wars" by Michael T Klare • **see** also "Guns, Germs and Steel" by Jared Diamond • **see** also "Straw Dogs" by John Gray 2002
• **Tools for Peace** • **see** "Preparing for Peace" lectures • examples taken from "War Prevention Works" Dylan Mathews Oxford Research Group 2001• "The Peace Book" by Louise Diamond 2001.
• **Inequality** • diagram from UNDP Human Development Report 1999 • US prison population: Washington Post February 2000 • US health cover: The Economist 20 January 2001 • figures on poverty: UNICEF Report September 2001 • "Understanding Capitalism" by Douglas Dowd
• **Interest-free Banking** • **see** also "Time Banks: No More Throw-Away People" by Edgar S Cahn 2000
• **Wörgl** • **see** also "After the Black Death" by George Huppert 1986 • The "Chiemgauer" is a voucher local currency system started in Germany in 2002 and attracting much attention from other communities, **see** www.Chiemgauer-regional.de
• **Global Eco-Currency** • Douthwaite's strategy is selected because it relates the monetary system to environmental and societal needs with great clarity • **see** also Center for Concern factsheet on "Rethinking Bretton Woods." • "Beyond Globalization" by Hazel Henderson.

• Citizens' Dividend • **see** www.progress.org • "Economic Justice for All" by Michael Murray 1997 • BIEN Basic Income Network Europe. ww.bien.be • The Tragedy of the Commons - destroyed by *uncontrolled* use. Under market fundamentalism global commons are made available to companies for commercial profit. They should be regulated for fair use by all. **See** various discussion groups on the web • "The Natural Wealth of Nations" by David Roodman.

• Basic Needs • Max-Neef's *matrix of needs and satisfiers* is more useful for development studies **than** Maslow's *hierarchy of needs.*

• "Them or Us?" • **see** "An Intelligent Person's Guide to the Classics" by Peter Jones 1999 • real democracy has been very rare, perhaps only practiced by Athenians from 507 to 322BC and, of religious groups, by the Quakers. It was attempted by the American Founders, particularly Thomas Jefferson. However democracy is the basis of much indigenous culture and was the basis of Gandhi's thought • Lathur is a project of ASSEFA using Gandhian techniques in Indian villages.

• Wealth in Poverty • **see** Amartya Sen's work on comparative wealth • for hunter-gatherer attitudes see "Ishmael" and "Beyond Civilization" by Daniel Quinn.

• Just Change • ACCORD at Gudalur. TN. India is also working with Oxfam • Latrine workers see "Endless Filth" by Mari Marcel-Thekaekara.

• Free Trade • **see** Center for Concern factsheet on Corporate Accountability • Banana dispute see Rachel's E&HW No. 679 • For policy recommendations see "Hungry for Trade" p121 • Analysis of free trade damage

see "Winners and Losers" The Economist April 28, 2001 • "Passion for Free Markets" articles by Noam Chomsky on www.zmag.org.

• WTO • **see** "Another World is Possible" by Fisher & Ponniah 2003 • **see** CNES websites to "demystify the World Bank, IMF and WTO, particularly how they diminish the role of government and facilitate privatization of service delivery often without the knowledge or consent of citizens or elected officials." • "International Trade & Sustainable Development" an Earthscan reader VA.

• Water Denied • many positive local books like "Rainwater Harvesting" by Shree Padre (India) • "Resource Wars" by Vandana Shiva 2002.

• Ending Tyranny • **see** also Aliance for Democracy websites to "end corporate domination, and achieve true democracy, a just society and sustainable and equitable economy." • "The Post-Corporate World" by David C Korten 1999.

• The Empire • "Full Spectrum Dominance" is the key phrase in Joint Vision 2020 by US Department of Defense 2001 and is used as a mission statement by all branches of US military • "The West and the Rest" by Roger Scruton 2002 • "Al Qaeda" by John Gray 2003

• Imperial Tribute • "Why are the Americans Smiling?" Hugo Salinas Price www.plata.com.mx • Dr Thomas Bennett, professor at the US Naval War College, published by the US Naval Institute January 2002 pp53-56.

• Soil • Phosphorus is a limited resource, **see** Folke Gunther, Feasta Review 2001 and www.holon.se/folke

• Feeding the World • **see** Center for Concern factsheet on Food Security • US NAS quote from "Transgenic Plants and World Agriculture" July 2000.

• **Organics** • Yields: Rodale Institute in Kutztown, Pennsylvania 2001 • Cuba: "Cuba's Organic Revolution" by Walter Schwarz. Resurgence 212 • mineral content: www.foodisyourbestmedicine.com.

• **Agrichemical Revolution** • We will only confuse kids if we continue to refer to the Green Revolution, green now having a totally different connotation • US NAS on GM danger, reported in Rachel issue 695 • Time Magazine, Jan 99, found 81% of Americans want labelling of food content and origin, and by the end of that year New York Times "biotech industry poll" found 93% want labelling. However the powerful industry lobby has so far prevented this • "An Agricultural Testament" by Sir Albert Howard 1940.

• **Citizens' Jury** • The Jefferson Center has prepared a short booklet describing the Citizens Jury process in detail • for official report of the procceedings in Andhra Pradesh **see** International Institute for Environment and Development: www.iied.org **and** Institute of Development Studies: www.ids.ac.uk

• **POPs & EDCs** • **see** also "Pandora's Poison" by J Thornton.

• **Microbes** • Edward Hooper, in "The River", gives compelling evidence that the HIV virus was transferred from chimpanzees to humans by a biotech company experimenting with a polio vaccine called Chat in Zaire in 1957-60. The company tested Chat on more than a million Africans though the "patients" did not need a vaccination. A polio epidemic occurred soon after in the vicinity, probably caused by the vaccine. **See** also article by Matt Ridley in Prospect magazine June 2000 • **see** Die Off article on Disease.

• **Genes** • Genome destabilization of rats: Research at University of Leicester, New Scientist 11 May 2002 • "Genome" by Matt Ridley 1999.

• **Making New Plants** • "Seeds of Doubt" by the Soil Association 2002 is based on inspections and interviews with farmers in the US and Canada. The facts have, to a great extent, been suppressed in the US by massively funded financial pressure from the biotech industry, so research from this UK organization is significant • Allergenic GM: US NAS 2000 report on biotech foods • The Impact of GM on Agriculture, Food and Health by the BMA (UK) 1999 • Transfer of toxins to soil: Saxena in NATURE Dec 1999 • The nutritional value of GM vitamin A rice is a cynical PR exercise by biotech companies. Indian NGOs like SCAD in Tirunelveli (scad@md4.vsnl.net.in) prevent deficiency with a diet of yellow fruit and vegetables (papaya, pumpkin, etc) which grow like weeds at no cost. They say that education is necessary, and not novel seeds that deplete genetic diversity and make poor people dependent on bought seed.

• **Commercial Eugenics** • "Maybe One" by Bill McKibben 1998.

• **Commercial Eugenics** • **see** also the work of professor Steven Rose, Brain & Behaviour Research Group, Oxford University i.e. "The sanguine conceit that somehow we have the power to stop or direct evolution is the most arrogant fantasy of all."

index

glossary

C&C	Contraction and Convergence, 37-9	
DTQ	Domestic Tradable Quota, 40	
DU	Depleted Uranium, 48, 50	
EDCs	endocrine-disrupting chemicals, 131,136,142-43,153	
EPA	US Environmental Protection Agency, 119	
EU	European Union, 13	
FTAA	Free Trade Area of the Americas, 110	
G8, G21, G40		
	Group of rich, medium, poor countries, 39, 88, 97	
GCI	Global Commons Institute, 37	
GATT	General Agreement on Tariffs and Trade, 104	
GATS	General Agreement on Trade in Services, 107-12	
GHG	greenhouse gases, 25, 30, 33, 38-39, 46, 74, 99	
GE, GM, GMO		
	genetic engineering, 101,131,142,146,149,152-56,159	
IMF	International Monetary Fund, 97	
IPCC	Intergovernmental Panel on Climate Change, 27-9, 31, 33-4, 36	
LETS	Local Exchange Trading Schemes, 75, 172	
NAFTA	North American Free Trade Argreement, 20, 110, 118	
NAS	US National Academy of Sciences. 29	
NBIC	Nano-Bio-Info-Cogno, 169	
PNAC	Project for the New American Century, 120	
POPs	Persistant Organic Pollutants, 131, 136, 142-44, 153	
TRIPS	Trade Related Intellectual Property Rights, 63, 106	
UNDP	United Nations Development Program, 61	
USDA	US Department of Agriculture 135, 157	
WTO	World Trade Organization, 97-8, 104-08, 113, 118, 144, 153, 173	

WHY DO PEOPLE HATE AMERICA?

By Ziauddin Sardar
&
Merryl Wyn Davies

Hardcover &
Trade Paperback

240 Pages

Trade
Paperback

$12.95

ISBN
0-9713942-5-3

All royalties and profits to UNICEF

LINES IN THE SAND

New Writing on War and Peace

Edited by Mary Hoffman and Rhiannon Lassiter

Trade Paperback

Juvenile Fiction / Short Stories

Ages 8+

288 Pages

$7.95

ISBN
0-9729529-1-8

Must Read.

What do CNN, your history teacher, and the White House have in common?

ABUSE YOUR ILLUSIONS
Edited by Russ Kick
Oversized Softcover, 352 Pages
ISBN 0-9713942-4-5

50 THINGS YOU'RE NOT SUPPOSED TO KNOW
By Russ Kick
Trade Paperback, 128 Pages
$9.95 • ISBN 0-9713942-8-8

The Disinformation Guide series edited by Russ Kick has become the definitive place to find revelations about government cover-ups, scientific scams, corporate crimes, medical malfeasance, historical whitewashes, media manipulation, and other knock-your-socks-off secrets and lies.

www.disinfo.com

notes